IET TELECOMMUNICATIONS SERIES 67

Managing the Internet of Things

Other volumes in this series:

Volume 9	**Phase noise in signal sources** W.P. Robins
Volume 12	**Spread spectrum in communications** R. Skaug and J.F. Hjelmstad
Volume 13	**Advanced signal processing** D.J. Creasey (Editor)
Volume 19	**Telecommunications traffic, tariffs and costs** R.E. Farr
Volume 20	**An introduction to satellite communications** D.I. Dalgleish
Volume 26	**Common-channel signalling** R.J. Manterfield
Volume 28	**Very small aperture terminals (VSATs)** J.L. Everett (Editor)
Volume 29	**ATM: the broadband telecommunications solution** L.G. Cuthbert and J.C. Sapanel
Volume 31	**Data communications and networks, 3rd edition** R.L. Brewster (Editor)
Volume 32	**Analogue optical fibre communications** B. Wilson, Z. Ghassemlooy and I.Z. Darwazeh (Editors)
Volume 33	**Modern personal radio systems** R.C.V. Macario (Editor)
Volume 34	**Digital broadcasting** P. Dambacher
Volume 35	**Principles of performance engineering for telecommunication and information systems** M. Ghanbari, C.J. Hughes, M.C. Sinclair and J.P. Eade
Volume 36	**Telecommunication networks, 2nd edition** J.E. Flood (Editor)
Volume 37	**Optical communication receiver design** S.B. Alexander
Volume 38	**Satellite communication systems, 3rd edition** B.G. Evans (Editor)
Volume 40	**Spread spectrum in mobile communication** O. Berg, T. Berg, J.F. Hjelmstad, S. Haavik and R. Skaug
Volume 41	**World telecommunications economics** J.J. Wheatley
Volume 43	**Telecommunications signalling** R.J. Manterfield
Volume 44	**Digital signal filtering, analysis and restoration** J. Jan
Volume 45	**Radio spectrum management, 2nd edition** D.J. Withers
Volume 46	**Intelligent networks: principles and applications** J.R. Anderson
Volume 47	**Local access network technologies** P. France
Volume 48	**Telecommunications quality of service management** A.P. Oodan (Editor)
Volume 49	**Standard codecs: image compression to advanced video coding** M. Ghanbari
Volume 50	**Telecommunications regulation** J. Buckley
Volume 51	**Security for mobility** C. Mitchell (Editor)
Volume 52	**Understanding telecommunications networks** A. Valdar
Volume 53	**Video compression systems: from first principles to concatenated codecs** A. Bock
Volume 54	**Standard Codecs: image compression to advanced video coding, 3rd edition** M. Ghanbari
Volume 59	**Dynamic Ad Hoc Networks** H. Rashvand and H. Chao (Editors)
Volume 60	**Understanding Telecommunications Business** A Valdar and I Morfett
Volume 65	**Advances in Body-Centric Wireless Communication: Applications and State-of-the-art** Q. H. Abbasi, M. U. Rehman, K. Qaraqe and A. Alomainy (Editors)
Volume 68	**Advanced Relay Technologies in Next Generation Wireless Communications** I. Krikidis and G. Zheng
Volume 905	**ISDN applications in education and training** R. Mason and P.D. Bacsich

Managing the Internet of Things

Architectures, Theories and Applications

Edited by
Jun Huang and Kun Hua

The Institution of Engineering and Technology

Published by The Institution of Engineering and Technology, London, United Kingdom

The Institution of Engineering and Technology is registered as a Charity in England & Wales (no. 211014) and Scotland (no. SC038698).

First published 2016

The Institution of Engineering and Technology
Michael Faraday House
Six Hills Way, Stevenage
Herts, SG1 2AY, United Kingdom

www.theiet.org

British Library Cataloguing in Publication Data
A catalogue record for this product is available from the British Library

ISBN 978-1-78561-028-8 (hardback)
ISBN 978-1-78561-029-5 (PDF)

Typeset in India by MPS Limited
Printed in the UK by CPI Group (UK) Ltd, Croydon

Contents

Chapter 1

Topology control for building scalable energy-efficient Internet of Things

Jun Huang[1]*, Qiang Duan*[2] *and Cong-cong Xing*[3]

Abstract

Internet of Things (IoT) is one of the most significant recent developments in the field of networking. Energy efficiency and scalability are two important technical issues that must be fully addressed in order to build high-performance IoTs. Topology control (TC) plays a crucial role in scalable and energy-efficient IoTs. In this chapter, we first give an overview of TC technologies and their applications in IoTs. Then we describe a systematic approach for topology construction in IoT to achieve scalability and energy efficiency. Such an approach includes a hierarchical system framework for IoT deployment, an optimization model for realizing energy efficiency, and an algorithm for solving the optimization mode. Experimental results are also presented in the chapter to show effectiveness of this topology construction approach.

1.1 Introduction

Internet of Things (IoT) as an emerging concept has gained worldwide attention from both industries and research communities. The term IoT is composed of two keywords – Internet and Things – and their definitions shape the ultimate form of an IoT. First, "Internet" here indicates that IoT is based on the existing Internet infrastructures and designed to perform data transmission and information exchange in a rather global and pervasive manner, instead of "Intranets" that only process data within its local range [1,2]. Second, associated with the widest range, the term Things is supposed to have the broadest pattern, that is, all uniquely identifiable computing devices are potential IoT elements. Note that equipment as actuator or radio transceiver is optional for a legitimate IoT element – RFID and items labeled by

[1]Institute of Electronic Information and Networking, Chongqing University of Posts and Telecommunications
[2]Department of Information Science and Technology, The Pennsylvania State University
[3]Department of Mathematics/Computer Science, Nicholls State University

various scanning-compatible codes could of course be a part of IoT, even they provide no extra capability other than a unique identifier.

With rapid adoption of IoT in various fields, diverse requirements have risen from both academic and industrial communities for this emerging technology. Especially at present, the connectivity density of Internet is at a dramatic scale, as the amount of interconnected devices is increasing in booming fashion. On the other hand, most of the "things" interconnected within IoT are devices with constrained resources, especially limited battery capacity. Therefore building large-scale energy-efficient IoT is particularly important. We summarize the following three characteristics as the most vital requirements for IoT.

Perception capability: This capability, via a variety of sensors or identifiers, etc., essentially provides an approach to linking real objects (i.e. with measurable physical quantities) with digitalized and Internet-compatible data. In the big picture of an IoT eco-system, all these elements for perception work jointly as the access from where they need to exist to the place where people want them to present on the Internet.

Transmission quality: The tremendous advance of telecommunication and networking industry brings not only neat transmission efficiency and greatly diverse wireless pathways, but also a highly intertwined communication scenario. IoT naturally has a good reason to employ the state-of-the-art underlying data transmission technologies in its backbone such as the fourth Generation Long-Term Evolution (LTE) and its several variants approved by third Generation Partnership Project (3GPP) organization. However, data transmission in IoT faces some new challenges due to some special features of this new networking paradigm, such as high node density, *ad hoc* network architecture, low network bandwidth among nodes, and limited battery capacity at regular nodes.

Data processing ability: Nodes in an IoT should be able to make intelligent decisions on how to cope with the collected data. With the aid of Internet access, this processing can be supported by Cloud computing or other remote acceleration strategies, but an IoT device/node should be equipped with the least computing resource to accomplish the very first step of data processing.

The requirements of IoT bring some challenges for its implementation. Differing from traditional wireless sensor networks (WSNs), IoT achieves a larger scale and becomes more complex. In addition, an IoT typically consists of more objects that consume higher power, which requires that energy efficiency to be considered seriously.

Topology control (TC), in general, refers to a group of techniques used in distributed computing and networking systems to alter underlying network topology to enhance system performance and/or reduce system cost. Lately, TC technologies have been shaped as a loop divided into construction phase and maintain phase of a networks connectivity [3]. Conventionally, TC is used mostly in autonomous, infrastructure-free wireless *ad hoc*, sensor networks and their variants. With advances on these network formations, a great number of literatures [4,5] have been published and some theoretical or practical solutions based on graph theory are well grounded already. Comparatively, Internet is infrastructure based and has its own more fixed topology, thus are often absent of TC. With advancement of IoT, which combines the traditional infrastructure-based Internet backbone with infrastructure-free *ad hoc* sensor networks, TC technologies have been explored as an effective solution to build large-scale energy-efficient IoT.

The backbone part of IoT is basically an Internet and therefore, follows highly hierarchical Internet topology under conventional infrastructure. On the other hand, the microstructure or local topology of an IoT may include *ad hoc* networks in which TC may be applied for enhancing system performance. It is reasonable to see this atomic element as a small network, in which its scale could still be a few hundreds or even thousands of nodes but ignorable compared with the massive scale of a complete IoT that includes the backbone Internet infrastructure. This is to say that the overall topology of IoT will not affect the local network upon analysis, and vice versa. Consequently, the process of this atomization and isolation opens up a window of bridging the gap of classical TC techniques and cutting-edge perspective of IoT architecture.

IoT is globally device-heterogeneous but locally homogeneous. It can be assumed that network architecture is a device-similar or homogeneous in a smaller region or in a single applications coverage. Unlike in wired networks where power-supply is basically regarded unlimited, devices in IoT are wireless and autonomous and thus are mostly battery-operated. This property makes emphasis on a key optimization objective C energy-efficiency – for large-scale IoT.

Mobility is concerned in IoT but not a primary focus. Relatively speaking, Wireless Sensor Network (WSN) tends to be stationary after initial deployment while Mobile Ad hoc NETwork (MANET) has higher mobility. Currently, applications such as military surveillance, tag tracking, intelligent transportation control, and smart home do not involve much mobility C they basically exchange information at one location.

In the rest of this chapter, we will first give an overview of TC techniques available for IoT and then present a hierarchical framework of topology construction that facilitates building scalable and energy-efficient IoT. We formulate topology construction with this framework as an optimization problem and give an algorithm for determining the number and location of relay nodes in this framework to achieve energy efficiency.

1.2 Overview of TC in IoT

In this section, we give an overview of existing TC technologies. TC can be defined as an iterative loop combining initialization phase, topology construction phase, and topology maintain phase. Initially, all nodes in network discover their nearby nodes with maximum transmission power to establish an initial topology C a connectivity graph. In the construction phase, network control employs some algorithms to build a reduced topology, where nodes are operated under an adjusted lower transmission power. As battery consumption fails some nodes and nodes may have moved over time, maintain phase takes its place and triggers new topology construction when necessary. As we discussed in the previous section, the local IoT architecture tends to have low mobility and thus is rather static. Therefore, our overview of TC in this section focuses on the topology construction phase.

Given the interpretation of topology construction, the algorithms for building reduced topology (connectivity graph) suggest the very criteria of any TC technique. These algorithms are designed with different optimization objectives. We list here three most common objectives and their corresponding applicable circumstances.

Energy Preservation: Since most of the devices such as sensors and actuators connected in IoT have limited battery capacity, a priority concern for IoT design should be conservation of energy. Otherwise, overly short life-time of nodes will dramatically damage the value in use for an IoT ecosystem. Lower energy consumption may be achieved with both hardware- and software-based approaches. Recent advancements of microelectronic technologies, improved hardware design of IoT devices has significantly reduced energy consumption. Given the same set of devices and connectivity restrictions, network control software may employ some algorithms to create an appropriate topology that has lower energy consumption thus long network life time. This is the fundamental motivation of TC. Various algorithms have been developed based on graph theory to form a network topology and reduce the total number of links needed for keeping all active communication paths in IoT while maintaining connectivity for each node.

Interference/Collision Avoidance: Another optimization objective for TC techniques is the interference among proximate radio-equipped devices, since occurrence of interference/collision imposes a huge negative impact on devices interrelate data transmission. Interference tends to be more severe in networks with higher node density. For this reason, the initialization phase with maximum transmission power and densest connectivity graph has the highest probability of severe interference. That is why the reduced topology generated by TC algorithms in construction phase could relieve this problem to some degree.

Capacity/Throughput Improvement: Some specific interference scenes cause another destructive phenomenon. It occurs when a node is prevented from sending packets to other nodes due to a neighboring transmitter. In some literatures, this phenomenon is said to be Hidden Node Problem. Some TC algorithms are designed with an objective to prevent node placement that causes such a problem thus improving network capacity.

Overall, the above three objectives must be thoroughly considered in TC in order to achieve high-performance IoT. Please note that misuse of TC will degrade or even eliminate positive effect. Therefore selection or design of TC algorithms must take into account compatibility with IoT under particular scenarios.

1.3 A framework of topology construction for scalable energy-efficient IoT

In this section, we present a framework of topology construction for building a large-scale energy-efficient IoT.

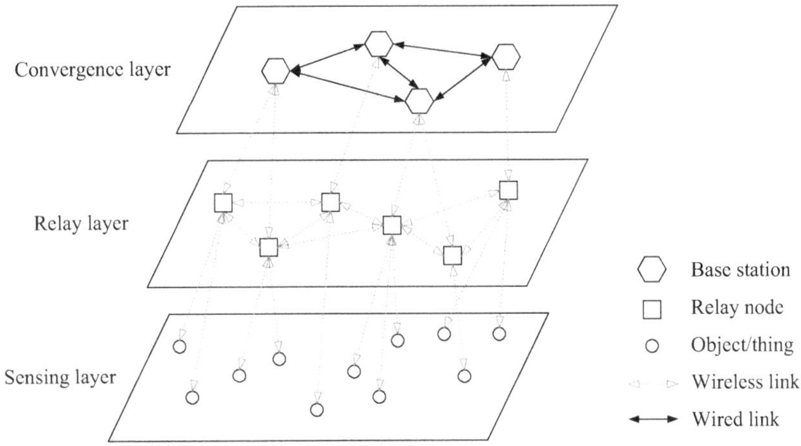

Figure 1.1 An example of system framework for IoT deployment

Previous engineering practice for deploying large-scale WSNs shows that dynamic routing is not very effective in wide area outdoor environments. Many factors such as electromagnetic interference, air humidity, and temperature, all have great impacts on sensor's data transmission, thus making such a network structure ineffective for large-scale networking. More importantly, WSNs configured with dynamic routing protocols require network nodes to perform power-consuming data processing for path computation. The power consumption of data processing increases significantly with network scale due to the complexity of dynamic routing protocols for a large number of nodes. In addition, dynamic routing requires network nodes to exchange route information among them regularly, which not only causes overhead traffic in network but also consumes node power for extra data communications. All these factors make such a network deployment scheme ineffective for building scalable energy-efficient IoT. On the other hand, network elements deployed in IoT very often have low mobility and network topology remains relatively stable, which makes dynamic routing gain little advantage over static routing configuration.

Based on these observation and consideration, we argue that for large-scale IoT it is reasonable to adopt static routing for higher power efficiency in order to achieve better network scalability and longer life time. We present a tiered framework for IoT topology construction in which objects are placed in a hierarchical structure with static routing configuration. As shown in Figure 1.1, this framework includes three layers – Sensing layer, Relay layer, and Convergence layer from bottom to up. Sensing layer is used for placing objects and things (e.g., RFID), Relay layer is formed by a collection of relay nodes, and Convergence layer consists of several base stations that are connected to the Internet. With the purpose of energy saving and link load balancing, the objects/things (sensing nodes) in the sensing layer are not allowed to communicate with each other directly. Instead, the communication between any two objects must go through a relay node. That is, equipments in sensing layer can only send data to a relay node in the upper layer. On the other hand, the sensing node receives few signaling packets from relay nodes. Such signaling packets are quite small, which can be neglected compared with data sending from sensing node. Nodes in the Relay layer form a relay network where any two neighbor nodes can communicate with each other. The other major functionality of a relay node is to forward the data from the sensing layer to a base station in the upper layer. In the convergence layer, base stations are also interconnected to form a network, which further uploads data to the Internet.

Note that this framework provides a general scheme for IoT topology construction to achieve scalability and energy-efficient. This framework is not limited to any specific technical implementation thus is applicable for various application scenarios. By placing IoT elements in the above hierarchical framework, this sensor deployment scheme provides flexibility, promotes scalability, and promises increased manageability. One of the major benefits introduced by such a tiered paradigm is that equipments in IoT do not require sophisticated hardware and do not need to run complex routing mechanisms and thereby significantly reduce the network cost.

1.4 Modeling topology construction for scalable energy-efficient IoT

In order to enable the tiered scheme of IoT topology construction presented in the last section, we formulate the system framework as follows.

The communication policy for the tiered topology construction scheme includes the following key aspects:

1. No sensor node can communicate with any other sensor node, even the distance between a pair of sensor nodes is less than the communication radius.
2. A sensor node can communicate with a relay node if the distance between these two nodes is less than the communication radius.
3. Two relay nodes may communicate with each other if the distance between them is less than the communication radius.

In addition, we make the following assumptions about the system framework:

- All the nodes in the framework are in the fixed sites.
- Nodes in the same type have the same attribute, e.g. initial energy, energy consumption parameters, maximum sending power, minimal receiving power, and so forth.
- Nodes in the Sensing layer can send data to a base station in a multi-hop manner.

- Each node in both Sensing and Relay layer is energy-constrained, while base station is not.
- The whole network of IoT represents a connected graph, that is, each node in the Sensing layer has a path to a base station, so does the relay node.

Given such hierarchical system framework, the goal of topology construction is to determine the number and location of relay nodes while satisfying power-saving and budget constraints. In this section, we define the system constraints according formulate the topology construction as an optimization problem. Listed following are notations of variables and parameters used in this chapter.

E_{tx}, E_{rx}: the energy consumption at a node for data transmission and receiving, respectively.

E_{elec}: the energy consumption of radio electronics.

ϵ_0, ϵ_1, ϵ_2: transmit amplifier of the node, sensing node, and relay node, respectively.

d_{ij}: the distance between node i and node j.

L: the data length.

F_{ij}: the data rate from node i to node j.

F_{max}: the maximum data rate of a link.

$C_{\mathbb{S}}$, $C_{\mathbb{R}}$, $C_{\mathbb{B}}$: the monetary cost of sensing node, relay node, and base station.

$|\cdot|$: the cardinality of a set.

l, m, n: the cardinality of set $\mathbb{S}, \mathbb{R}, \mathbb{B}$

W_0: the system budget.

Denote the entire network of IoT as $G(N, A)$, where N represents the node set and A represents the wireless link set. A sensing node can only communicate with a relay node in the upper layer, whereas a relay node can send/receive data both to/from its neighbor relay nodes as well as a base station; therefore, $G(N, A)$ is a directed and connected graph. We call node i and node j neighbors if i and j are able to communicate with each other. Let $\mathcal{N}(i)$ be the set of i's neighbors, $\mathcal{C}$ be the adjacency matrix of $G(N, A)$, then:

$$\mathcal{C} = \begin{bmatrix} c_{11} & c_{12} & \cdots & c_{1|N|} \\ c_{21} & c_{22} & \cdots & c_{2|N|} \\ \vdots & \vdots & \ddots & \vdots \\ c_{|N|1} & c_{|N|2} & \cdots & c_{|N||N|} \end{bmatrix} \tag{1.1}$$

where $c_{ij} = 1$ if $j \in \mathcal{N}(i)$, otherwise $c_{ij} = 0$.

We consider the following system constraints.

Energy Consumption Constraints: From the system perspective, the energy consumption of IoT mainly comes from data communication because the energy expenditure in data sensing and processing is much less compared to data communication [6]. Thus, only the energy consumption of data communication, i.e. energy for sending and receiving data, is taken into account in this model. According to the Friis free space model [7], we have:

$$E_{tx} = (E_{elec} + \epsilon_0 \cdot d^2) \cdot L, \quad E_{rx} = E_{elec} \cdot L \tag{1.2}$$

Upon above two equations, the data length L from node i to node j in a time unit is equal to the data rate from i to j. Therefore, the energy consumption per time unit of each node can be calculated by:

$$e_i = \sum_{j \in \mathbb{R}} c_{ij} \cdot F_{ij} \cdot (E^{\mathbb{S}}_{elec} + \epsilon_1 \cdot d^2_{ij}) \qquad \forall i \in \mathbb{S} \tag{1.3}$$

$$e_j = \sum_{i \in \mathbb{S} \cup \mathbb{R}} c_{ij} \cdot F_{ij} \cdot E^{\mathbb{R}}_{elec} + \sum_{i \in \mathbb{B} \cup \mathbb{R}} c_{ji} \cdot F_{ji} \cdot (E^{\mathbb{R}}_{elec} + \epsilon_2 \cdot d^2_{ji}) \qquad \forall j \in \mathbb{R} \tag{1.4}$$

$$e_k = \sum_{j \in \mathbb{R}} c_{jk} \cdot F_{jk} \cdot E^{\mathbb{B}}_{elec} \qquad \forall k \in \mathbb{B} \tag{1.5}$$

where e_i, e_j, and e_k denote the consumption of sensing node, relay node, and base station. $E^{\mathbb{S}}_{elec}$, $E^{\mathbb{R}}_{elec}$, and $E^{\mathbb{B}}_{elec}$ are the energy consumption of radio electronics of sensing node, relay node, and base station, respectively.

Link Flow Balancing Constraints: In the IoT, the base stations are usually interconnected by wired links, which have wider bandwidth compared with relay and sensing nodes, therefore, the bandwidth is constrained at nodes except for base stations. For a relay node, it communicates with not only its neighbor relay nodes but also sensing nodes in the lower layer. Thus, the wireless links of a relay node should satisfy:

$$c_{ij} \cdot F_{ij} + c_{ji} \cdot F_{ji} \leq F_{\max} \qquad \forall i, j \in \mathbb{R} \tag{1.6}$$

Likewise, the wireless links at each sensing node and base station need to meet the constraint $c_{ij} \cdot F_{ij} \leq F_{\max}$, where $\forall i \in \mathbb{S}, j \in \mathbb{R}$ or $\forall i \in \mathbb{R}, j \in \mathbb{B}$.

System Budget Constraint: Since relay nodes and base stations are comparatively expensive, the deployment of an IoT must be as cheap as possible. On the other hand, the number of base stations is fixed, consequently, the IoT deployment should meet the system budget constraint, i.e.:

$$0 < C_{\mathbb{S}} \cdot l + C_{\mathbb{R}} \cdot m < W_0. \tag{1.7}$$

1.5 Topology construction algorithm for scalable energy-efficient IoT

The main purpose of this chapter is to reduce energy consumption to achieve an energy-efficient IoT. Hence, the optimization model for energy-efficient IoT deployment is defined as:

$$\begin{aligned}
&\min\left[\sum_{i\in\mathbb{S}} e_i + \sum_{j\in\mathbb{R}} e_j + \sum_{k\in\mathbb{B}} e_k\right]\\
&\text{s.t.}\\
&e_i = \sum_{j\in\mathbb{R}} c_{ij}\cdot F_{ij}\cdot (E_{elec}^{\mathbb{S}} + \epsilon_1\cdot d_{ij}^2) \qquad \forall i\in\mathbb{S}\\
&e_j = \sum_{i\in\mathbb{S}\cup\mathbb{R}} c_{ij}\cdot F_{ij}\cdot E_{elec}^{\mathbb{R}} + \sum_{i\in\mathbb{B}\cup\mathbb{R}} c_{ji}\cdot F_{ji}\cdot (E_{elec}^{\mathbb{R}} + \epsilon_2\cdot d_{ji}^2) \qquad \forall j\in\mathbb{R}\\
&e_k = \sum_{j\in\mathbb{R}} c_{jk}\cdot F_{jk}\cdot E_{elec}^{\mathbb{B}} \qquad \forall k\in\mathbb{B}\\
&c_{ij}\cdot F_{ij} + c_{ji}\cdot F_{ji} \le F_{\max} \qquad \forall i,j\in\mathbb{R}\\
&c_{ij}\cdot F_{ij} \le F_{\max} \qquad \forall i\in\mathbb{S}, j\in\mathbb{R}\ or\ \forall i\in\mathbb{R}, j\in\mathbb{B}\\
&0 < C_{\mathbb{S}}\cdot l + C_{\mathbb{R}}\cdot m + < W_0
\end{aligned} \tag{1.8}$$

We devise a Minimal Energy Consumption Algorithm (MECA) as shown in Algorithm 1 in order to solve the problem (10). The basic idea behind MECA is to first apply canonical K-means clustering algorithm to select the relays, then construct a graph to associate each edge a weight through mapping the transmitting/receiving energy of the connected node pair. At last, MECA employs a well-known *Steiner* tree algorithm to solve the problem. Specifically, MECA works in the following four steps.

In the first step (line 1), MECA applies K-means clustering algorithm [8] in the sensing layer to find a set of clusters, and then the closest relay from each cluster is selected to form the set $\mathbb{R}_1$, where K-means clustering is a method of cluster analysis which aims to partition n observations into K clusters in which each observation belongs to the cluster with the nearest mean. The set $\mathbb{R}_1$ is essentially the minimal single-cover set due to the minimal clusters found by K-means.

The second step (lines 2–8) establishes the graph connecting relays and base station. It also assigns the weight on each edge.

In the third step of MECA (line 9), it employs a well-known *Steiner* tree algorithm [9] to compute a minimal energy consumption tree. Note that the *Steiner* tree algorithm used in this chapter is similar to that used in Reference 10. The major difference is that the MECA exploits the energy consumption as the link weight instead of the weight as defined in Reference 10. In such a way, the number and location of relay nodes can be determined by the *Steiner* tree algorithm, which guarantees the entire network to be energy-efficient.

The forth step obtains the solution min(e) for optimization problem (10) by summing the total weight on each edge, where:

$$\min(e) = \min\left[\sum_{i\in\mathbb{S}} e_i + \sum_{j\in\mathbb{R}} e_j + \sum_{k\in\mathbb{B}} e_k\right] \tag{1.9}$$

the total weight here means the multiples of the weight on the common links, which transmits data to base stations for two or more sensing nodes.

1.6 Performance evaluation

In this section, we give performance evaluation of the presented topology construction scheme through numerical experiments. The nodes in each topology are distributed in a $100 \times 100\,\mathrm{m}^2$ region. The number of candidate locations for placing relay node is 99, and $m = 1$. The parameters are configured as follows. We set $E_{elec} = 50\,\mathrm{nJ/bit}$, $E^{\mathbb{B}}_{elec} = 2E^{\mathbb{R}}_{elec} = 4E^{\mathbb{S}}_{elec} = 4E_{elec}$, $\epsilon_1 = \epsilon_2 = 100\,\mathrm{pJ/bit/m^2}$, $F_{ij} = 100\,\mathrm{kbps}$ for sensing nodes, $F_{ij} = 200\,\mathrm{kbps}$ for relay nodes, and $F_{\max} = 400\,\mathrm{kbps}$.

Figure 1.2 shows the number of relay nodes deployed to achieve energy-efficient IoT in the different communication radius. In this experiment, the number of sensing

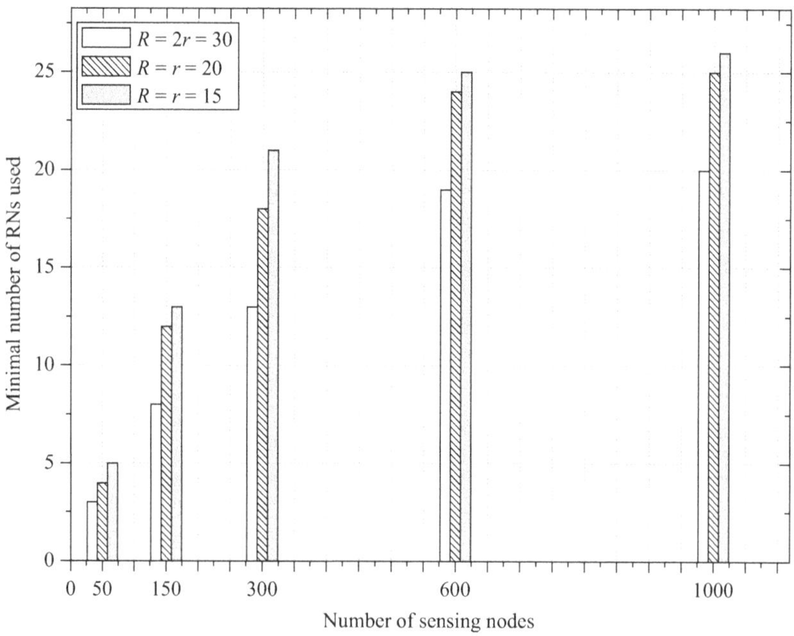

Figure 1.2 Number of relay nodes used for five topologies with different communication radius

nodes was set to be 50, 150, 300, 600, and 1 000 to represent the different scale of IoT. From the figure we can see that, the minimal number of relay nodes increases with network scale. This is natural because the larger the network scale is, the more relay nodes are needed for covering all the sensing nodes. Another interesting observation we obtained from the figure is that when the number and positions of sensing nodes and base stations are stable, the minimal number of relay nodes decreases as the communication radius of relay nodes increases. This is because larger communication radius allows a relay node to cover more sensing nodes; thus reducing the number of relays required for deploying. In addition, we also find that a small variation in communication radius has little impact on the minimal number of relay nodes. In particular, Figure 1.2 gives similar minimum numbers of relay nodes for the cases of 600 and 1 000 nodes. Since the nodes are randomly distributed in a $100 \times 100\,\text{m}^2$ region, the nodes density is relatively high. Therefore, the minimal number of relay nodes for IoT deployment would be affected by not only communication radius but also the node density.

Figures 1.3–1.6 give the relationship between energy consumption of deployed IoT and the number of relay nodes with different communication radius. It can be seen from these figures that the network energy consumption per unit time increases with the number of sensing nodes. Specifically, Figure 1.4 shows the energy consumption for another topology where $l = 150$, $n = 1$. We observe that the energy consumption tends to be high while the communication radius of relay nodes increases. It turns

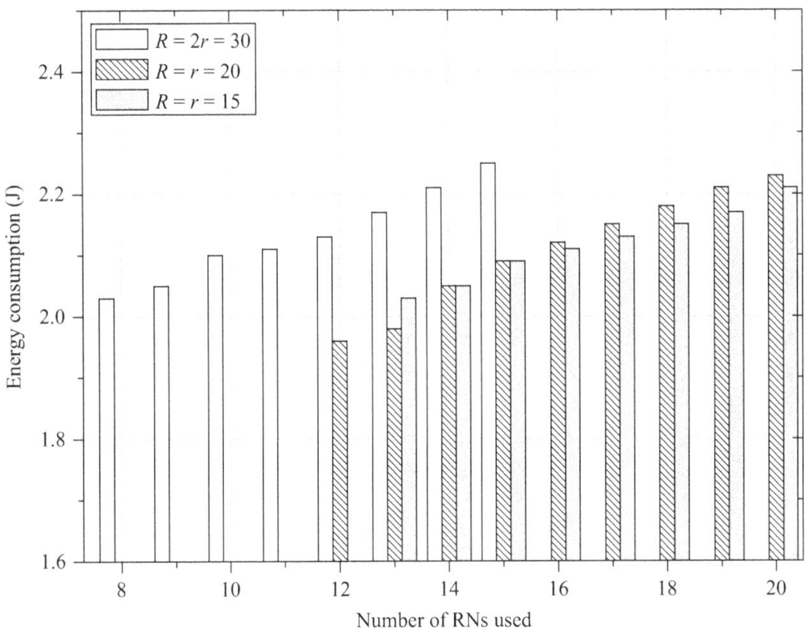

Figure 1.3 Energy consumption of deployed IoT ($l = 150$) versus the number of relay nodes with different communication radius

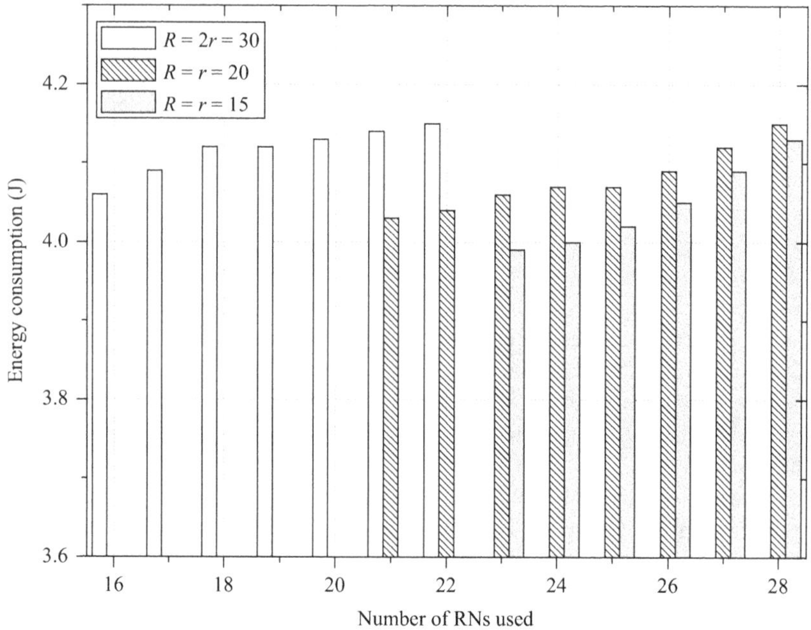

Figure 1.4 Energy consumption of deployed IoT ($l = 300$) versus the number of relay nodes with different communication radius

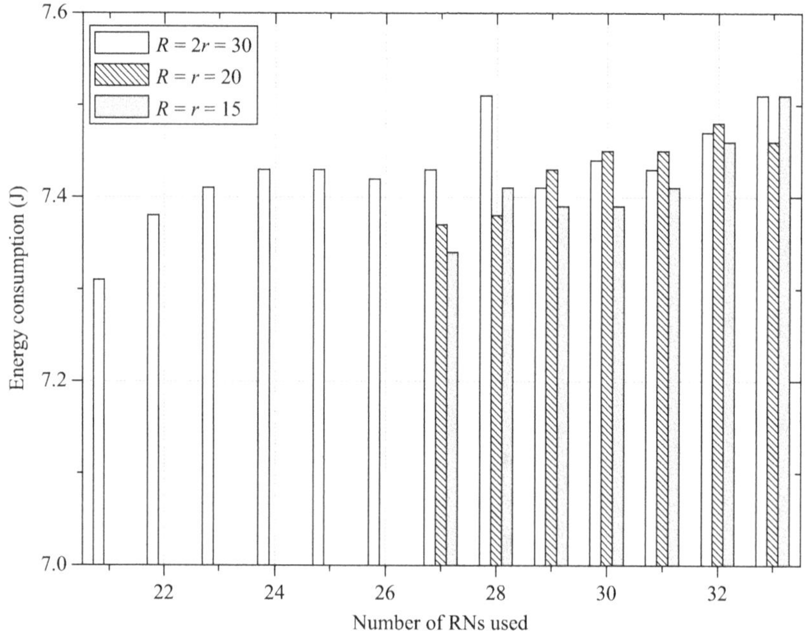

Figure 1.5 Energy consumption of deployed IoT ($l = 600$) versus the number of relay nodes with different communication radius

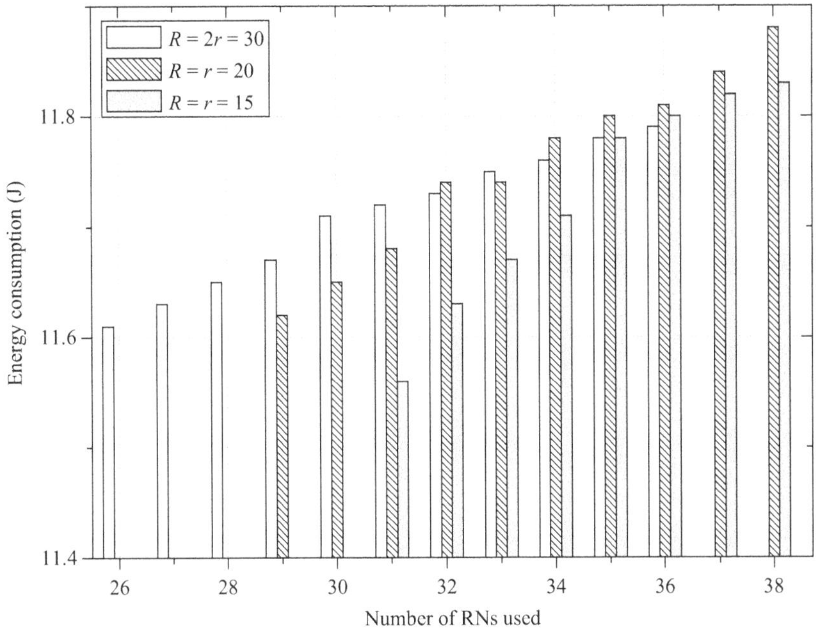

Figure 1.6 Energy consumption of deployed IoT (l = 1 000) versus the number of relay nodes with different communication radius

out that the minimal number of relay used in this topology is reached when $R = 2$ and $r = 30$. In addition, we can also see that the IoT is unconnected when $R = r = 20$ or $R = r = 15$, and $m < 12$ or $m < 13$, which leads to zero network energy consumption. Note that we assume the network energy consumption be zero if the graph is unconnected. Since the goal of deploying a energy-efficient networked IoT is to place as few relay nodes as possible in the IoT, we only need to consider the energy consumption for the scenario with the least relay used. Therefore, it is not necessary to consider the case of $m > 15, R = 2r = 30$, and we assume that the energy consumption is also zero in this setting. Figures 1.5–1.7 give the data of energy consumption for the network when $l = 300$, $l = 600$, and $l = 1\,000$, respectively. These figures provide the same insight as Figure 1.4 does, which proves the proposed algorithm robustness.

In order to validate the effectiveness of our proposed hierarchical deployment structure, we implement a hybrid deployment scheme for comparing the network lifetime of the two frameworks. The main difference between the hybrid scheme and our proposed hierarchical framework is that the sensor nodes in the lower layer of the hybrid structure are allowed to communicate directly with their neighboring nodes. Figures 1.7–1.9 show the comparing results of network lifetime for different numbers of nodes with various communication ranges.

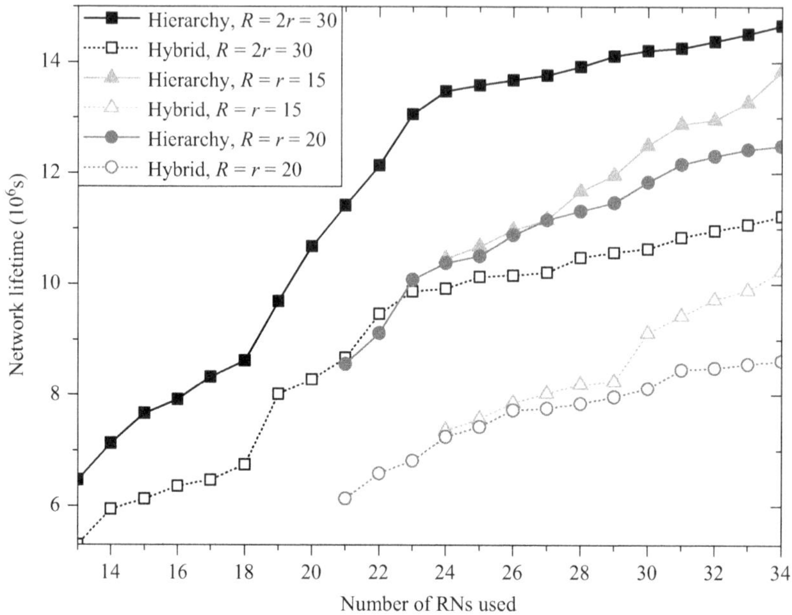

Figure 1.7 Network lifetime comparisons for the topology $l = 300, m = 83, n = 1$

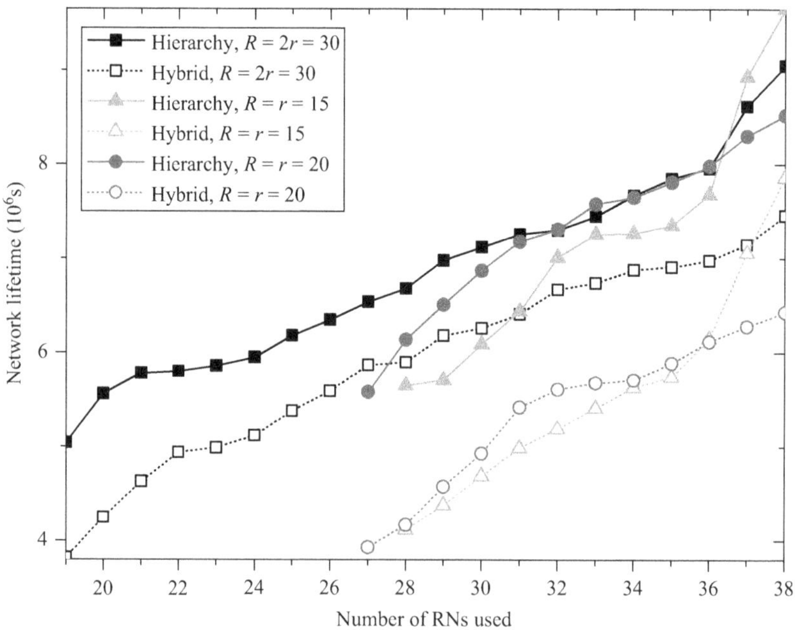

Figure 1.8 Network lifetime comparisons for the topology $l = 600, m = 111, n = 1$

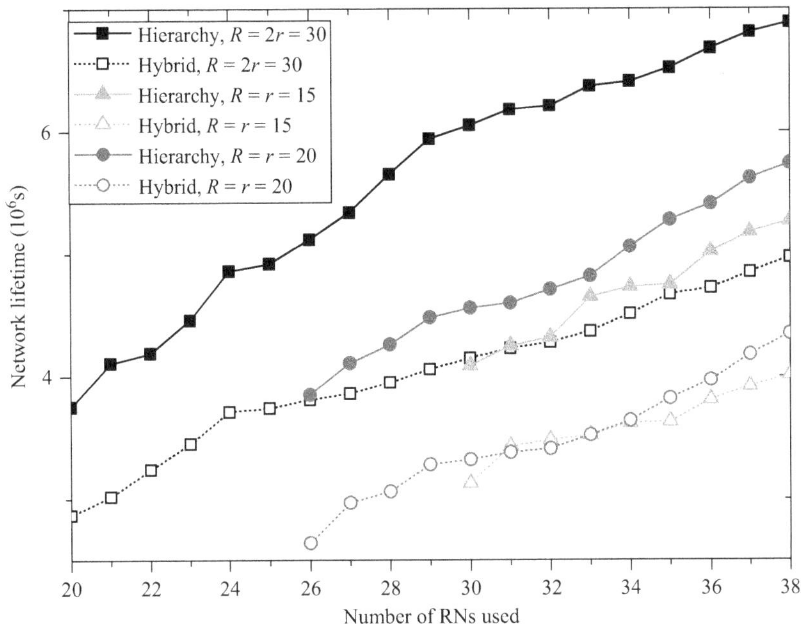

Figure 1.9 Network lifetime comparisons for the topology $l = 1\,000, m = 111, n = 1$

From these figures, we find that the network lifetime of IoT deployed in the hierarchical structure is longer than that of IoT deployed in the hybrid scheme. This is due to the fact that in the hybrid scheme the sensing nodes near to relay nodes may be overloaded, therefore, consume more energy than other nodes, which causes a shorter network lifetime. While in the hierarchical scheme, sensing nodes send information to their neighbor relay nodes, then the relay nodes forward such information to a base station. In this way, the hierarchical scheme balances the network load of the nodes; thus balancing node energy consumption and extending the network lifetime. The more relay nodes are deployed, the longer the network lifetime is. In addition, increasing communication radius of relay nodes would also prolong the network lifetime. Another important fact we can see from these figures is that the network lifetime becomes shorter when the number of sensing nodes increases. Since more sensing nodes result in more network traffic, which leads to a shorter network lifetime, the proposed hierarchical deployment scheme is better than the hybrid scheme with respect to the network lifetime. Therefore, we claim that the proposed hierarchy scheme is more preferable for energy-efficient deployment of IoT.

1.7 Conclusions

In this chapter, we discussed TC technologies and their applications in Internet of Things for achieving scalability and energy efficiency, which are two critical technical issues for building high-performance IoTs. We first gave an overview of TC technologies and how they can be applied in IoTs for improving network scalability and energy efficiency. Then we describe a topology construction scheme for large-scale energy-efficient IoTs. This scheme includes a hierarchical system framework for network deployment, an optimization model for minimizing network energy consumption, and an algorithm to solve the optimization model. We show through numerical experiment results that the presented scheme can achieve longer network lifetime in large-scale IoTs.

References

[1] L. Atzori, A. Iera, and G. Morabito, "The Internet of Things: A survey," *Computer Networks,* vol. 54, no. 15, pp. 2787–2805, 2010.

[2] D. Miorandi, S. Sicari, F. De Pellegrini, and I. Chlamtac, "Internet of things: Vision, applications and research challenges," *Ad Hoc Networks,* vol. 10, no. 7, pp. 1497–1516, 2012.

[3] Y. Wang, "Topology control for wireless sensor networks," in *Wireless Sensor Networks and Applications.* Berlin, Germany, Springer, 2008, pp. 113–147.

[4] P. Santi, "Topology control in wireless ad hoc and sensor networks," *ACM Computing Surveys (CSUR),* vol. 37, no. 2, pp. 164–194, 2005.

[5] R. Rajaraman, "Topology control and routing in ad hoc networks: A survey," *ACM SIGACT News,* vol. 33, no. 2, pp. 60–73, 2002.

[6] I. Akyildiz, W. Su, Y. Sankarasubramaniam, and E. Cayirci, "Wireless sensor networks: a survey," *Computer Networks,* vol. 38, no. 4, pp. 393–422, 2002.

[7] W. Heinzelman, A. Chandrakasan, and H. Balakrishnan, "An application-specific protocol architecture for wireless microsensor networks," *IEEE Transactions on Wireless Communications,* vol. 1, no. 4, pp. 660–670, 2002.

[8] M. Li, M. Ng, Y.-m. Cheung, and J. Huang, "Agglomerative fuzzy k-means clustering algorithm with selection of number of clusters," *IEEE Transactions on Knowledge and Data Engineering,* vol. 20, no. 11, pp. 1519–1534, 2008.

[9] L. T. Kou, G. Markowsky, and L. Berman, "A Fast Algorithm for Steiner Trees," *Acta Informatica,* vol. 15, no. 2, pp. 141–145, 1981.

[10] D. Yang, S. Misra, X. Fang, G. Xue, and J. Zhang, "Two-tiered constrained relay node placement in wireless sensor networks: Computational complexity and efficient approximations," *IEEE Transactions on Mobile Computing,* vol. 11, no. 8, pp. 1399–1411, 2012.

Algorithm 1 MECA

Input:
$\mathbb{S}, \mathbb{R}, \mathbb{B}, R \geq r > 0$

Output:
Minimal Energy Consumption $\min(e)$

Apply K-means clustering algorithm to obtain a single-cover set $\mathbb{S}_1 \subseteq \mathbb{S}$, choose the closest relay $i \in \mathbb{R}$ to replace the $j \in \mathbb{S}_1$ forming the set $\mathbb{R}_1$.
for $i \in \mathbb{R}, j \in \mathbb{R} \cup \mathbb{B}, i \neq j$ **do**
  Calculate the distance d_{ij} between i and j;
  if $d_{ij} \leq R$ **then**
    Add the node i and j to a candidate set RN for placement, set $c_{ij} = 1$ in G;
  end if
end for
Assign edge weight for G in terms of (4), (5) and (6) on each edge;
Apply a well-known *Steiner* Tree algorithm to compute a minimal energy consumption *Steiner* tree G^T of $G = (\mathbb{S} \cup RN \cup \mathbb{B}, A)$ spanning the node set $\mathbb{B} \cup \mathbb{R}_1$.
for each edge in G^T **do**
  Sum the total weight on each edge, denoted as $\min(e)$;
end for
return $\min(e)$;

Chapter 2
Wireless sensor network operating systems: a survey

Haiying Zhou[1,2], Xing Liu[2,5], Shen Lin[3], Jian Li[4], Shengwu Xiong[2] and Kun-Mean Hou[5]

Abstract

Operating system (OS) is a critical research topic in the wireless sensor network (WSN). With an outstanding WSN OS, not only the constrained WSN platform resources can be managed efficiently, but also the complicated WSN application development can be simplified soundly. In this chapter, a survey to the current WSN OSes is investigated. Different OS concerns such as the OS architecture, the scheduling model, the memory management, the application programming, the application reprogramming, the energy conservation, the real-time scheduling, and the fault tolerance are reviewed. In addition, the features of the different OSes are compared and the ongoing research challenges are proposed. The work presented in this chapter can be helpful for the WSN users to select an appropriate OS for the WSN motes, and it can also be useful for the WSN developers to set forth the future OS design directions.

2.1 Introduction

The Internet of Things (IoT) is comprised of the "things" objects which are embedded with the sensors, the electronics, the software, and the network connectivity. With the environmental data collection and the data exchange among these intelligent

[1]School of Electrical and Information Engineering, HuBei University of Automotive Technology, 442002 Shiyan, HuBei, China
[2]Hubei Key Laboratory of Transportation Internet of Things, Wuhan University of Technology, 430070 Wuhan, China
[3]Key Laboratory of Automobile Parts Testing, General Administration of Quality Supervision, Inspection and Quarantine of P.R. China, 441003 XiangYang, HuBei, China
[4]School of Computer Science, Harbin Institute of Technology, 150001 Harbin, China
[5]LIMOS Laboratory, UMR 6158 CNRS, Blaise Pascal University, Les Cézeaux, BP 10125, 63173 Clermont-Ferrand, France

IoT objects, the integration between the physical world and the computer-based systems can be realized [1–3].

In recent years, many research works have been researched in the IoT, and these works can be classified three visions: the thing-oriented vision, the Internet-oriented vision, and the semantic-oriented vision [3]. In the thing-oriented field, the WSN is one of the critical research issues. With the WSN technique, the physical or environmental data can be collected and forwarded to the common network framework, e.g., the Internet network. And then, these data can be analyzed and used for the smart decision. Currently, the WSN technique has played a significant role in the proliferation of the IoT paradigm [4–6].

As the proliferation of the WSN technique, many research challenges emerge and need to be addressed. First, the WSN motes are typically the small-size and low-cost ones on which the platform resources such as the memory resource and the energy resource are constrained [5,6]. Therefore, the memory optimization and the energy conservation become essential to the WSN. Second, the WSN hardware and software platforms are currently diverse [7]. Different hardware platforms such as the BTnode and the iMote2, and the software OSes such as the TinyOS, the Contiki, and the SOS, are developed. Due to this reason, the users need to understand the diverse low-level platform details to make the programming and this process is hard for many users. Third, many WSN motes are deployed in the outdoor harsh contexts where the labor maintenance to the deployed motes is quite difficult. Thus, the fault-tolerant ability and the remote reprogramming become important to the WSN motes.

To address the challenges above, the research and implementation of an outstanding WSN OS becomes significant. With an outstanding OS, not only the constrained WSN platform resources can be managed efficiently, but also the complicate WSN application development can be simplified soundly. Due to this significance, a survey to the current popular WSN OSes is investigated in this chapter. Different OS concerns are discussed and the OS ongoing research challenges are proposed. This work presented in this chapter can be helpful for the WSN users to understand the different WSN OSes quickly, and it can also be useful for the WSN developers to set forth the future WSN OS design directions.

The structure of this chapter is organized as follows: From section 2.2 to section 2.9, the features of different WSN OSes, such as the architecture model, the scheduling model, the memory management, the application programming, the application reprogramming, the energy conservation, the real-time performance, and the fault recovery mechanisms, are investigated respectively. In section 2.10, the comparison and the research challenges of the current WSN OSes are summarized.

2.2 OS architecture

OS architecture can have an effect not only on the kernel code size but also on the way how the system services are provided. The current WSN OS architectures can be classified into three types: the monolithic architecture, the modular architecture, and the virtual machine (VM) architecture.

2.2.1 Monolithic architecture

In the monolithic-architecture OSes, all the software components are built together to form a single system image. This kind of OS has compact code size and sound execution efficiency. Yet, the entire software binary needs to be updated if any change to the software program is made. As a result, the application reprogramming performance of this kind of OSes is low. Currently, the monolithic architecture has been used in the OSs such as the TinyOS [8], the openWSN [9], and the uCOS [10].

2.2.2 Modular architecture

Modular architecture is different from the monolithic architecture in that parts of the software components, such as the applications and the drivers, can be built independently into a set of dynamic loadable modules (Figure 2.1). Currently, this kind of OS architecture has been used in the SOS [11].

In the modular-architecture OSes, a module's address is changeable after it is updated, thus the position independent code (PIC) and the indirect interaction mechanisms are commonly applied in these OSes. With the PIC mechanism, the interactions within a module can still keep valid even if the module changes its address. With the indirect interaction mechanism, the interactions among the modules can still keep valid even if the modules locates to a new address. In Figure 2.1, the indirect interaction mechanisms of the SOS are depicted.

The advantage of the module-architecture OS is that the modules can be updated dynamically without reprogramming the monolithic software binary. Therefore, the application reprogramming or the system fault repairing can be performed efficiently. Yet, the module management such as the module loading, module deletion, and module registration needs to be performed. As a result, the OS code size and the architecture complexity will increase. Moreover, the modules need to interact with each other in the indirect way, and the OS execution efficiency will be decreased by this.

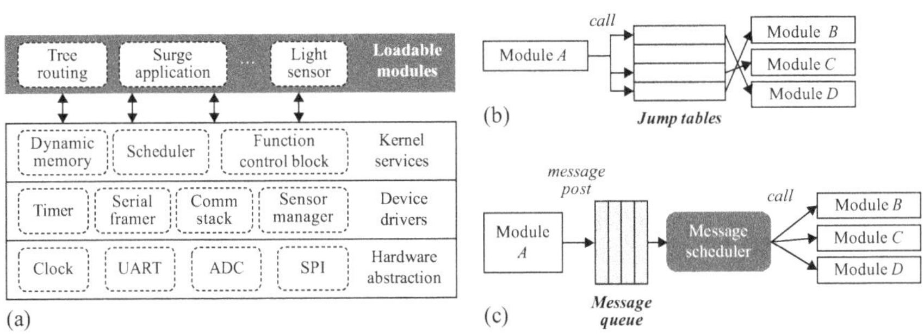

Figure 2.1 (a) Software structure of the SOS. Modules can be loaded dynamically. (b, c) Module interaction mechanism in the SOS. Modules interact with each other indirectly either by the synchronous jump table or the asynchronous message queues

2.2.3 VM architecture

In the VM-architecture OSes, the application code is decoupled from the low-level system code and built independently into an interpretable image. During the run-time, the application image can be updated dynamically and be interpreted by the VM.

A typical example of the VM architecture WSN system is the embedded Java VM (EJVM). With the EJVM, the users can program the WSN nodes by the popular, reusable, robust, object-orient, and hardware-independent Java language. By this means, a user-friendly application development environment can be provided to the WSN users. Currently, the VM architecture has been applied in many WSN OSes, including the simpleRTJ [12] (Figure 2.2), the LeJOS [13], the nanoVM [14], the JavaCard [15], and the DarjeelingVM [16].

Since the VM application code is executed in the interpreted way, the application of the VM-architecture OSes can be hardware independent, and this is significant for the programming in the heterogeneous WSN contexts. In addition, the VM application code is decoupled from the low system code. Thus, only the application part needs to be updated during the reprogramming process, and the reprogramming performance keeps high.

Yet, the code execution efficiency of the VM-architecture OSes is low since the application code is executed by interpretation, and this causes more energy to be cost during the run-time process by comparing with that of the native application code. In addition, the memory cost of most VM-architecture OSes is high since the byte code interpreter and the byte code handlers need to be implemented, e.g., about 20 KB code memory is required by the simpleRTJ to run on the 8-bit AVR microcontroller ATmega1281, and this makes these EJVMs not suitable to run on the high memory-constrained WSN nodes.

2.2.4 Discussion

Different architectures have different advantages and drawbacks, and they are appropriate to be applied in different contexts. The monolithic TinyOS has low memory

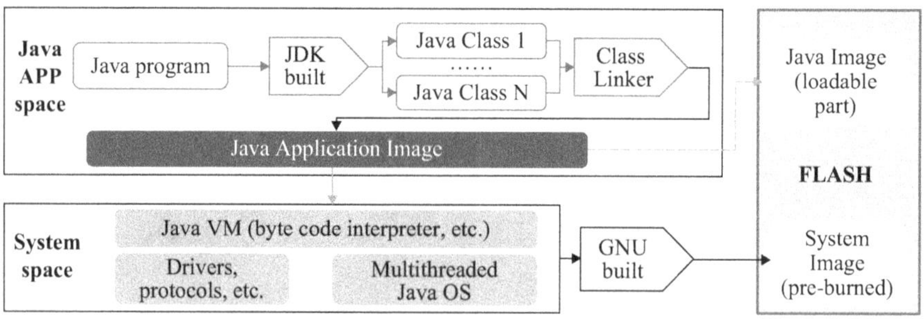

Figure 2.2 Elementary diagram of the Java VM/OS SimpleRTJ. SimpleRTJ is consisted of the embedded JVM and the lightweight multithreaded JavaOS. With the JavaOS, the WSN applications can be programmed by the multithreading Java language

and energy cost, yet the application reprogramming performance is poor, thus it is suitable to be used when the platform resource is highly constrained and the application reprogramming is rarely performed. The modular SOS can dynamically load both the application module and the system service modules. Thus, it is essential to be used when the application reprogramming or the system fault repairing is regularly needed. The VM-based simpleRTJ can provide a decent application development environment to the WSN users although its memory and energy costs are relatively high. Therefore, it can be a sound selection for the WSN developers in case the WSN platform resources are abundant.

2.3 OS scheduling model

The scheduling models of the WSN OSes include the event-driven scheduling, the preemptive multithreading, the cooperative multithreading, and the hybrid scheduling.

2.3.1 Event-driven scheduling and preemptive multithreading

The basic scheduling models of the WSN OSes include the event-driven scheduling and the preemptive multithreading [17,18]. The main difference between these two is that the latter supports the task preemption while the former cannot (Figure 2.3). With the preemption support, better real-time performance can be realized in the multithreaded OSes. Yet, the context switch needs to be performed during the preemption process. As a result, each task in multithreaded OS needs to be allocated an independent stack which will be used to store the context, and this causes the memory cost and the scheduling overhead of the multithreaded OSes to be higher than those of the event-driven OSes.

2.3.2 Cooperative multithreading

Cooperative multithreading is the scheduling mechanism in which the task currently controlling the CPU can yield the control to the others in the voluntary way, rather than in the preemptive way. It is differentiated with the preemptive multithreading

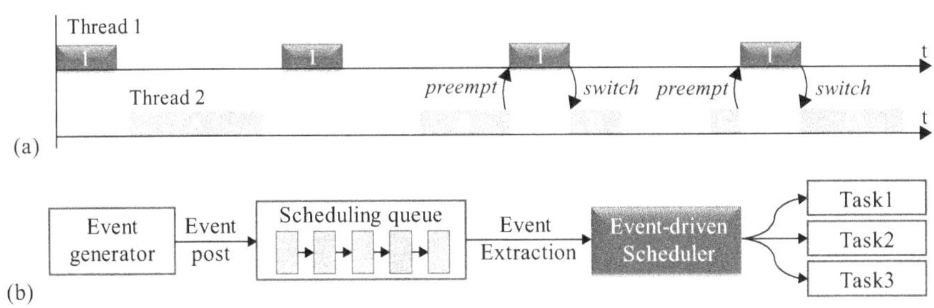

Figure 2.3 (a) Scheduling process of the multithreaded system. (b) Software structures of the event-driven system

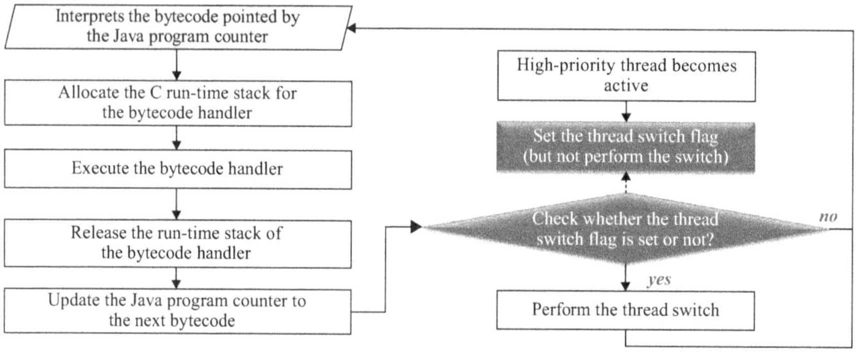

Figure 2.4 Cooperative scheduling model of the Java VM simpleRTJ. Only after the context of the current bytecode handler has been released, the thread switch can be performed. By so doing, the memory cost of the thread stacks can be optimized significantly

in which a high-priority task can preempt the low-priority task anytime outside of the low-priority task's control [19]. Currently, the cooperative multithreading has been used in several WSN OSes including the simpleRTJ and the Enix [20], and it has been used to reduce the stack memory usage and decrease the scheduling overhead.

In the Java VM/OS simpleRTJ, the cooperative multithreading mechanism, which is also termed as atomic instruction or bytecode-granularity thread switch in the JVM terminology, is applied to schedule the Java threads. With the cooperative multithreading, the thread switch will not be performed during the executing process of the bytecode handler, but it is performed until the handler runs to completion and the thread scheduler is called explicitly. Since the bytecode handler will not be preempted during the executing process, the allocation of an independent stack which will be used to store the run-time contexts of the handlers need not be performed. As a result, the stack memory cost of the simpleRTJ can be decreased significantly (Figure 2.4). In the Enix, the cooperative multithreading is also implemented. Yet, it is different from the simpleRTJ in that it is used to optimize the thread scheduling overhead and ease the thread race condition problem.

Although the cooperative multithreading can be used to optimize the memory cost and decrease the scheduling overhead, it still has some drawbacks. In the simpleRTJ, the memory cost is optimized, yet it is achieved at the cost of decreasing the OS real-time performance. In the Enix, the scheduling overhead is optimized, yet the software programming complexity is increased since the programmers need to be aware of calling the scheduler manually.

2.3.3 Hybrid scheduling

Hybrid scheduling is the scheduling mechanism in which the event-driven scheduling and the multithreading scheduling are both implemented. Currently, it has been applied in several WSN OSes including the TinyOS TOSThread [21], the Contiki [22],

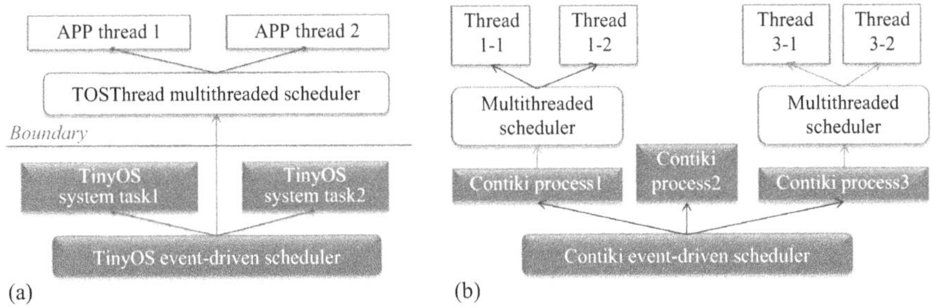

Figure 2.5 (a) Hybrid scheduling model in the TOSThread. (b) Hybrid scheduling model in the Contiki

and the HEROS [23], and it is designed with the purpose of simplifying the event-driven programming, adding the preemption into the event-driven system, or enabling the WSN OS to be context aware.

The hybrid scheduling structure of TinyOS TOSThread is depicted in Figure 2.5, the fully-preemptive thread package TOSThread is implemented on top of the TinyOS event-driven kernel, and it serves the TinyOS applications dedicatedly. With the TOSThread, the applications in the event-driven TinyOS can be programmed by the thread-based programming, and the programming complexity can be eased.

Similar to the TinyOS TOSThread, the multithreaded scheduler in the Contiki is also implemented on top of the event-driven scheduler. Yet, different from the TOSThread, the Contiki multithreaded scheduler can serve any Contiki task which requires the preemption support (Figure 2.5). By means of this multithreading extension, the long-running computations in the Contiki can be programmed while the time-sensitive nature of the Contiki can also be preserved.

In the HEROS, the multithreaded scheduler is also implemented on top of the event-driven scheduler. Yet, it is different from the TOSThread and Contiki in that the scheduling model is configurable. With the configurable functionality, the HEROS can run in three different models: event-driven model, multithreading model, and hybrid model (Figure 2.6). Since different models have different advantages and drawbacks, HEROS can become context aware and adapt well to the diverse WSN applications.

2.3.4 Implementation of different scheduling models

The event-driven scheduler is commonly implemented by using the event queue [24]. After an event is generated, it will be buffered in the event queue. The event scheduler polls this queue and extracts the event one by one. After an event is extracted, the corresponded task will be executed. Since the tasks cannot be preempted in the event-driven system, each task runs to completion before the next event can be extracted.

The implementation of the multithreaded scheduler includes the initialization of the thread stacks, the restoring of the thread contexts, and the selection of next thread to be scheduled. The thread selection depends on the scheduling strategy. Currently,

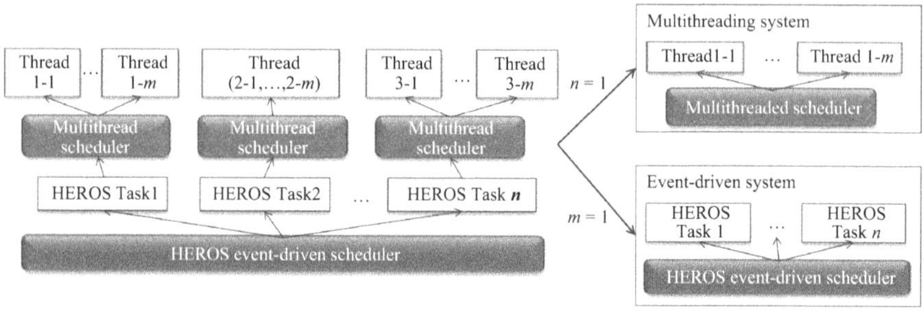

Figure 2.6 Hybrid scheduling model in the HEROS. The HEROS hybrid model can be configured either to be the event-driven model or to be the multithreaded model

the popular thread scheduling strategies in the WSN OSes include the Round-Robin (RR) scheduling, the Rate-Monotonic Scheduling (RMS), and the Earliest Deadline First (EDF) [25,26].

2.3.5 Discussion

Different scheduling models have different advantages and drawbacks, and they commonly strike the balance between the memory cost and the real-time performance. In case that the real-time responsiveness is required and the memory resource of the WSN platforms is abundant, the multithreaded scheduling can be considered. Reversely, the event-driven scheduling is more appropriate. Cooperative multithreading and the hybrid scheduling keeps the balance between the above two. As for which scheduling model to be selected, it depends on the application requirements and the WSN platform resources.

2.4 Memory management

Memory management is essential for the WSN since the RAM resources of most WSN nodes are constrained, e.g., the MicaZ node has only 4 KB RAM. To manage the memory resources efficiently, the dynamic allocation and the virtual memory mechanisms are popularly implemented in the WSN. In this section, the basic WSN dynamic allocation approaches, such as the SGS allocation and the SF allocation, are presented in section 2.4.1. Then, the extended dynamic allocation approaches, such as the coalescence-deferred sequential fit (SF) allocation and the defragmented SF allocation, are discussed in sections 2.4.2 and 2.4.3. Finally, the WSN virtual memory mechanisms are introduced in section 2.4.4.

2.4.1 Basic dynamic allocation mechanisms in the WSN

In Reference 27, the mechanisms of the dynamic allocation are classified into five kinds: the SF, the segregated free list which includes the simple segregated storage

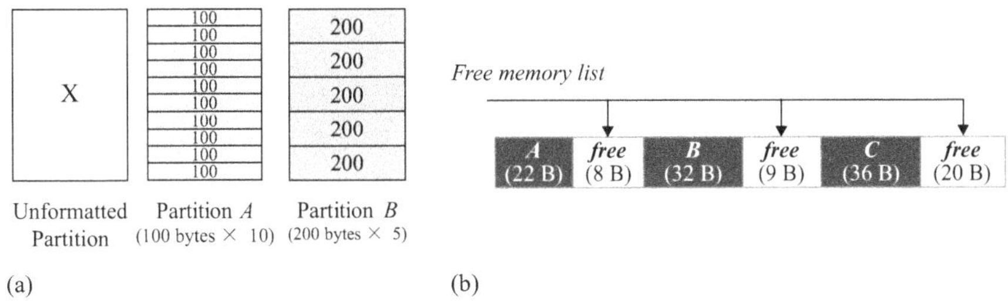

Figure 2.7 (a) SGS allocation in the uCOS. Partitions can be formatted during the run-time, and the different partitions can be formatted into the different fixed-size blocks. (b) SF allocation in the MantisOS. A free list is used to manage the free memory blocks

(SGS) and the segregated fit, the Buddy systems, the indexed fit, and the bitmapped fit. Currently, the SGS and the SF mechanisms are commonly used in the WSN OSs. The Buddy system approach is not popularly used in the WSN as the fragmentation percentage of this approach can be high [28]. The *malloc* allocation is not popularly used as nontrivial code space will be taken up by its implementation.

A key difference between the SGS and the SF is that the former performs the fixed-size block allocation while the latter performs the varied-size block allocation. In the SGS allocation, the memory heap is divided into segregated partitions. Each partition is formatted into a set of fixed-size blocks, and the block sizes in the different partitions are different (Figure 2.7). Since the fixed-size block allocation is performed, the allocation time of SGS is short and deterministic. Moreover, no block splitting and coalescence are needed. Yet, the internal and external memory fragmentation can be severe, and this reduces the memory utilization efficiency of this approach. Different from the SGS allocation, the block sizes of the SF allocators are not fixed. Each block is allocated in the sizes exactly as required. The advantage of this approach is that no internal memory fragments will exist. Yet, the allocation time is not deterministic as a free list needs to be searched upon allocation. Moreover, the block splitting and coalescence needs to be performed and this increases the allocation overhead (Figure 2.7). Currently, the SGS allocation has been applied in the Contiki [22] and the uCOS [10] while the SF allocation has been used in the MantisOS [29].

2.4.2 Coalescence-deferred SF allocation

In the SF system, block splitting and coalescence need to be performed, and this increases the allocation execution overhead. To optimize the allocation performance, the coalescence-deferred SF approach is proposed.

Coalescence-deferred mechanism is appropriate to be applied when the size of the newly allocated block is identical to that of the previously released block. In this case, the previous released block can be reused exactly for the allocation of the subsequent block, and it needs not to be coalesced. Currently, the coalescence-deferred SF has

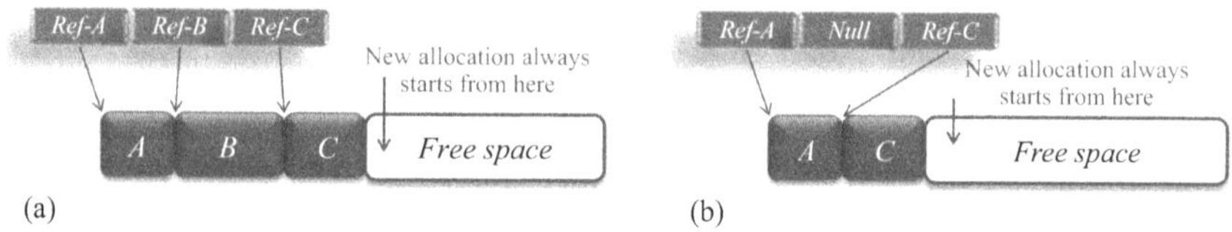

Figure 2.8 Contiki defragmented SF allocation. After the block B is released, the blocks A and C will be coalesced. By this means, the memory fragments can be reused. Since the addresses of some blocks will change after the coalescence, the indirect pointers needs to be used to access the blocks

been realized in the FreeRTOS [30]. With this mechanism, the allocation performance of the FreeRTOS can be optimized.

2.4.3 Defragmented SF allocation

In the SF allocation system, external memory fragment can occur, and this will reduce the memory utilization efficiency of this allocation approach, e.g., in Figure 2.7, if a 30-byte block needs to be allocated, it cannot be performed successfully although the total sizes of the free blocks are larger than 30 bytes. To utilize the external fragment efficiently, the memory defragmentation needs to be performed. Currently, this has been realized by the Contiki *mmem* allocator [31]. In the Contiki *mmem*, the fragments are defragmented proactively once it appears. In case that a block is released, the blocks adjacent to it will be coalesced (Figure 2.8). By doing this, all the free blocks will be continuous in the address and no memory fragments will exist.

2.4.4 Virtual memory mechanism

Virtual memory can support conceptually more memory than might be physically available, and makes the application programming easier by hiding the fragmentation of the physical memory. Typically, the implementation of the virtual memory requires the support from the hardware Memory Management Unit (MMU). Yet, most WSN microcontrollers are absent in the MMU hardware support. Therefore, the software-based virtual memory technique needs to be developed, and this has currently been realized in the Enix [20] and the t-kernel [32].

In the Enix, the software-based virtual memory mechanism is achieved by using the code insertion technique at the compiling time. The commonly used libraries are pre-built on the host PC and transformed into the PIC segments. Then, these libraries can be loaded dynamically into the code memory after being called by the user program.

In the t-kernel, the software-based virtual memory is also realized, and it is achieved by modifying the application instructions at the loading time. When a new page of application instructions needs to be executed, the corresponded instructions will be modified so as to achieve the virtual address translation. Currently in the t-kernel, a 64 KB virtual memory space over 4 KB physical memory can be supported.

With the virtual memory technique, the constrained memory resource of the WSN nodes can be utilized more efficiently.

2.5 Application programming model

Application programming model depends closely on the low-level OSes. In terms of the different OS scheduling models, the applications in the WSN can be programmed in two patterns: the event-based programming and the thread-based programming.

2.5.1 Event-based programming

Event-based programming is commonly applied in the event-driven OSes. In the event-driven systems, the preemption is not enabled, thus the execution time of the tasks cannot be lengthy. In order to program the lengthy task, the split-phase state-machine programming [33], with which the application code can be broken across multiple disjoint segments of code, is popularly used in the event-based programming. The advantage of this programming model is that the memory cost and the run-time overhead are low. The drawback is that the programmers need to construct the explicit machine states manually, and this is difficult to be handled by many developers.

2.5.2 Thread-based programming

Thread-based programming is commonly applied in the multithreaded OSes. It is different from the event-based programming in that the tasks can preempt each other. Thus, the complicated split-phase state-machine programming, which is needed in the event-driven programming, needs not to be performed in the thread-based programming. However, due to the thread switch operations, the memory cost and the run-time overhead of the thread programming are high, and this makes the thread-based programming less preferable for the high resource-constrained WSN devices. In addition, the thread synchronization operation, which can lead to the deadlock and the race condition problems, can also occur in the thread-based programming.

2.5.3 Thread-based programming in the event-driven OSes

The programming in the event-drive OSes is difficult to be handled since the programmers need to construct the machine states manually. To ease the programming complexity of the event-driven OSes, the user-friendly thread-based programming is motivated to be added into the event-driven OSes. Currently, this has been realized in several event-driven OSes, including the Contiki protothread [34] and the TinyOS TOSThread [21].

Contiki protothread does not support the thread preemption. Yet, it enables the users to program the event-driven Contiki applications by using the thread-based programming model. In the protothread system, a static local continuations (LC) variable will be defined in the task. This variable can be used to save and restore the execution address of the task. By doing this, the state-machine programming needs no more to be constructed and the programming complexity can be eased. TinyOS TOSThread is another mechanism which can support the thread-based programming in the event-driven TinyOS, and it is achieved by implementing a multithreaded scheduler on top of the event-driven scheduler. With the TOSThread, the users can program the TinyOS applications by the user-friendly thread-based programming.

2.6 Application reprogramming

Application reprogramming is practical and economic to be performed in the WSN since the application requirements and the network environments are changeable over the time. To avoid the work of recollecting the nodes by labor to make the reprogramming, the WSN reprogramming needs to be achieved through the wireless. However, the wireless communication is high energy cost [35] and the WSN bandwidth is commonly constrained [36–38]. Thus, the optimization to the reprogramming performance becomes essential in the WSN, and this can be achieved by reducing the reprogramming code size and developing a resource-efficient code dissemination protocol.

2.6.1 Optimization to the reprogramming code size

The optimization to the reprogramming code size can be achieved by the mechanisms such as the decoupling of the application code from the system code, the diff-patch approach and the code compression.

Once the application code is decoupled from the system code, only the application part needs to be updated during the reprogramming process. By doing this, the application reprogramming performance can be optimized. To decouple the application code from the system code, the module-architecture OSes (Contiki, SOS, etc.) or the virtual-machine architecture OSes (simpleRTJ, etc.) can be applied. In these OSes, only the small-size application code needs to be reprogrammed, and the reprogramming performance is high.

Diff-based approach can also optimize the reprogramming code size since only the differential changes between the new code and the old code needs to be updated by this approach. Currently, this approach has been used by several mechanisms such as the *Zephyr* [39], the *Hermes* [40], the *EasiLIR* [41], the *R*2 [42], and the Remote incremental update [43,44].

Code compression can optimize the code size as well [45], and it is appropriate to be applied in the WSN since the energy cost of decompressing the code can be lower than that of transmitting the large-size packet.

All the above approaches can be used comprehensively to optimize the reprogramming code size. With these approaches, both the reprogramming energy and the reprogramming time can be optimized significantly.

2.6.2 Code dissemination protocol

In addition to the reprogramming code size optimization, the development of a time-efficient and energy-efficient code dissemination protocol is also significant for the WSN reprogramming. Currently, different code dissemination protocols have been developed [46–50]. With these protocols, not only the reprogramming energy and time cost can be optimized, but also the reprogramming reliability and security can be improved.

2.7 Energy conservation

WSN nodes are prone to be deployed in the harsh environment where the power recharging is difficult or even impossible. In order to prolong the lifetime of the WSN network, the energy conservation becomes essential. Typically, a WSN node consists of the sensing subsystem, the processing subsystem and the communication subsystem, thus the energy conservation can be realized respectively from these aspects.

2.7.1 Energy conservation in the sensing subsystem

The sensing subsystem can be high energy cost in the WSN under many cases [51], e.g., the acoustic sensors [52] with the high-rate and high-resolution A/D converters can be power hungry during the sampling process. Thus, the energy conservation to the sensing subsystem becomes essential, and this can be achieved by the data prediction and the hierarchical sampling mechanisms.

Data prediction can conserve the energy cost by reducing the sensor sampling redundancy. It is feasible to be applied as some sensed phenomenon can be described by the given models. In case that the results predicted by these models are within acceptable error bounds, the sensing values queried by the users can be responded by sending the result calculated from this model, rather than the result sampled by the sensors. Currently, this technique has been achieved in several research works, e.g., the dynamic probabilistic model and the *Kalman* filter model have been applied respectively in References 53 and 54 to achieve the sensing data prediction.

The hierarchical sampling is another mechanism to conserve the sensing energy cost. With this approach, several types of sensors can be equipped on the nodes. These sensors have different resolutions and strike the tradeoff between the accuracy and the energy cost. During the run-time, one appropriate kind of sensors can be selected. For example, in the target tracking application, the low accurate sensors which has low energy cost can be selected to detect the targets. In case that the target objective is locked, the high accurate sensors which have higher energy cost will be loaded. By so doing, the sensor sampling process can be context aware and the energy can be utilized efficiently.

2.7.2 Energy conservation in the signal processing subsystem

It is experimented that the energy cost of transmitting one bit of sensing data is in a way the same as that of processing thousands of instructions on a typical WSN node [55]. Therefore, the optimization to the sensing data size is significant to achieve the energy conservation, and this can be realized by the data compression and data aggregation mechanisms.

Data compression is a common signal processing mechanism which can decrease the sensing packet size. It involves the operation of encoding the data at the sources and decoding it at the destination. Currently, many data compression approaches have been implemented in the WSN [56,57]. In addition to the data compression mechanism, the data aggregation [58,59], which can aggregate and fuse the sensing data from multiple other nodes so as to reduce the amount of the data transmitted to the base station, can also be applied to decrease the sensing data size. With this aggregation, the amount of the data traversing through the network can be reduced, and the energy cost can be optimized.

2.7.3 Energy conservation in the communication subsystem

Energy can also be conserved from the communication subsystem, and it is achieved either by reducing the communication redundancy, or by prolonging the inactive period of the radio transceiver.

Communication redundancy can exist in the WSN since the nodes are commonly deployed randomly in many cases and the number of deployed nodes can be larger than that is required. Thus, only partial nodes need to be selected to maintain the network while the others can fall asleep to conserve the energy. One typical example to reduce the communication redundancy is the topology control technique [60–63]. With this technique, only minimum numbers of nodes are selected to keep the active status of the network topology. By doing this, the network lifetime can be prolonged by a factor of two to three [63,64].

The optimization to the transceiver's active period is another approach to conserve the energy cost. Currently, it has been achieved by many protocols such as the TDMA-based protocols [65–67], the contention-based MAC protocols [68,69], and the energy-efficient routing protocols [70]. By using these protocols, the radio transceiver can be switched on only within a short period. Consequently, the energy cost can be optimized.

2.8 Real-time performance

Real-time performance is significant for the applications in which the nodes need to react immediately to the emergent events, such as the health care application, the gas leaking monitoring application, and the industrial engine control application. Currently, the real-time performance of the WSN can be improved from several aspects, including the task scheduling, the network protocol, and the data management.

From the task scheduling aspect, a real-time scheduling algorithm needs to be implemented to guarantee the deadline of the time-critical tasks [25,26]. Besides the real-time algorithms, the resource pre-reservation mechanism [71], with which the time-critical tasks can specify their resource demands and then the OS provides the timely and guaranteed resource access to these tasks, can also be used to improve the real-time performance of the WSN.

From the network protocol aspect, the delay-efficient MAC protocols [72,73] and routing protocols [74–76] can be developed. These protocols optimize the communication latency by applying the time-efficient scheduling mechanisms, easing the transmission confliction, prolonging the inactive period of the transceivers, easing the broadcast transmission, or selecting the optimal routing paths.

Data aggregation can also optimize the WSN real-time performance. It can ease the data redundancy which commonly exists due to the multiple sensor sampling and the slow change phenomena [58]. With the data aggregation, the network congestion, the communication and computational overhead can be decreased. Consequently, the real-time communication performance of the WSN can be enhanced. Besides the data aggregation, the data compression scheme, which can exploit the delta compression and provide better bandwidth utilization, can also be applied to optimize the communication latency [77].

2.9 Fault-tolerant mechanisms

The WSN network is prone to fall failure due to innumerous potential fault sources such as the unpredictable communication channel, the extreme outdoor environments, the battery depletion, as well as the fragility and the mobility of the nodes [78–80]. Since the sensor nodes can be deployed in the harsh environments where the post maintenance is hard, the fault tolerant mechanisms becomes significant for the WSN nodes. With these mechanisms, the WSN nodes can continue to function even if the faults occur.

Reference 81 classifies the WSN fault-recovery mechanisms into two types: active replication and passive replication. In the active replication system, all or many WSN nodes provide the same functionality. As a result, sufficient information can still be provided to the recipients even if parts of these nodes are failed, and the fault-tolerant ability of the WSN network can be improved. Currently, the active replication mechanism has been applied in the multipath routing [82] and the data aggregation [58] mechanisms. Different from the active replication approach, the nodes in the passive replication system are classified into the primary replicas and the backup replicas. In case that the primary replicas have failed, the backup replicas will be selected and continue providing the services. By doing this, the faults of the WSN can be recovered. One typical passive replication mechanism is the topology management mechanism [83]. With this mechanism, the network parameters such as the degree of network connectivity, the role of the nodes, the transmission power, can be adjusted during the run-time. With these operations, the network topology can be constructed and the network communication can be recovered from the fails.

2.10 Feature comparison and ongoing research challenges

2.10.1 Feature comparison of different WSN OSes

In Table 2.1, the features of different WSN OSes are compared. Each OS has the particular features and addresses the different WSN challenges. The SOS is a dynamic OS which can load the different modules dynamically. The RETOS is a robust

Table 2.1 Feature comparison of the different WSN OSes

WSN OSes	Architecture	Scheduling model	Memory management	Application program	Reprogramming part	Miscellaneous
TinyOS	Monolithic	Event-driven	Static allocation	Event, nesC language	Monolithic software image	TOSThread, TYMO, etc.
Contiki	Modular (only for application)	Hybrid	SGS allocation	Event & Thread, C language	Dynamic loadable application (by the relocation)	Coffee FS, protothread, etc.
SOS	Modular	Event-driven		Event, C	Dynamic loadable application (by using PIC code)	Module bug protection, garbage collection
MantisOS	Monolithic	Round-robin multithreaded	SF allocation	Thread, C	Monolithic software image	User-level stack, Remote shell
openWSN		Event-driven	Static allocation	Event, C		Full stacks (IEEE 802.15.4.e)
Enix	Modular (only for application part)	Cooperative/Priority-based/ Round-robin multithreading	Virtual memory	Thread, C	Dynamic loadable application (by using PIC code)	EcoFS, Reprogramming shell.
LiteOS		Priority-based/round-robin multithreading	Malloc at the kernel level	Thread, LiteC++	Dynamic loadable application (by differential patch)	UNIX-like LiteFS, LiteShell
nano-RK	Monolithic	Priority-based multithreading	Static allocation	Thread, C	Monolithic software image	Reservation Paradigm for timeliness
RETOS	Modular (only for application)	Round-robin/Event-aware multithreading	Single kernel stack and stack-size analysis	Thread, C	Dynamic loadable application (by relocation)	WSN-oriented network abstraction
t-kernel	Monolithic	Priority-based multithreading	Virtual memory	Thread, C	Monolithic software image	OS protection
simpleRTJ	VM	Cooperative/Round-robin multithreading	SF allocation & Garbage collection	Thread, Java language	Dynamic loadable application (by interpretation)	Garbage collection/ exception handling
uCOS	Monolithic	Priority-based/Round-robin multithreading	SGS allocation	Thread, C	Monolithic software image	Message passing, uC-GUI, uC-FAT, etc.
FreeRTOS		Cooperative/Priority-based multithreading	Coalescence-deferred SF/malloc			Run-time stack checking

and resilient OS which realizes the kernel extensibility with the dynamic reconfiguration [84]. The openWSN implements a fully open-source standards-based protocol stack. The LiteOS implements a UNIX-like file system [85]. And, the Enix implements a software-based virtual memory mechanism.

2.10.2 Research challenges of the WSN OSes

The different OS design and implementation mechanisms have addressed widespread WSN challenges effectively. Yet, some challenges still exist and need to be researched further. First, to achieve the real-time scheduling, the multithreaded scheduling needs to be implemented. Yet, the multithreaded systems have high memory cost and are not suitable to run on the tight memory-constrained WSN nodes. Thus, the research of a new scheduling mechanism which can achieve good real-time performance and yet keeps low memory cost becomes important. Second, the defragmented SF allocation which has been implemented in the Contiki can defragment the memory fragments so that the constrained memory resources can be utilized efficiently. Yet, the current Contiki allocator defragments the fragments proactively and the defragmentation overhead is high. Thus, the development of a new memory allocation mechanism which can defragment the fragments with low run-time overhead becomes essential. Third, different mechanisms have been implemented to conserve the energy cost of the WSN nodes, whereas the energy constraint is still a limitation for the proliferation of the WSN technology. Currently, the investigation of a new approach which can optimize the energy cost of the WSN platform further is still needed. Fourth, to reprogram the WSN nodes efficiently, it is significant to decouple the application code from the system code. The current dynamic loading or VM mechanism can achieve this objective, whereas the memory cost or energy cost of these mechanisms is high. Therefore, the development of a new memory-efficient and energy-efficient reprogramming scheme needs to be researched. Finally, the current WSN fault-tolerant mechanisms are effective to improve the availability of the network, but these approaches are limited in that they focus on the recovery of the network from the faults, and are lacking in the research of recovering the nodes from the faults. Therefore, the research of the new fault-recovery mechanism which can recover the WSN nodes from the faults becomes significant. In the ongoing works of the WSN OSes, the challenges above will be focused and addressed.

Acknowledgments

The authors would like to thank all the colleagues and copartners who have contributed to this study, and are also grateful for the research support from the Natural Science Foundation of HUAT (BK201411), the Sci-Tech. Pillar program of Hubei Province (2014BHE024), the National High-tech 863 R&D Program of China (2015AA015403), the fundamental research funds for the Central Universities (WUT:163110003, WUT:40120225), and the National Key Technology R&D Program (2012BAH45B01).

References

[1] Whitmore A., Agarwal A., Da Xu L. 'The Internet of Things—a survey of topics and trends'. *Information Systems Frontiers*. 2015;17(2):261–274.

[2] Miorandi D., Sicari S., De Pellegrini F., Chlamtac I. 'Internet of things: vision, applications and research challenges'. *Ad Hoc Networks*. 2012;10(7): 1497–1516.

[3] Atzori L., Iera A., Morabito G. 'The internet of things: a survey'. *Computer networks*. 2010;54(15):2787–2805.

[4] Borges L.M., Velez F.J., Lebres A.S. 'Survey on the characterization and classification of wireless sensor network applications'. *IEEE Communications Surveys and Tutorials*. 2014;16(4):1860–1890.

[5] El Emary I.M.M., Ramakrishnan S. *Wireless Sensor Networks: From Theory to Applications*. CRC Press; 2013.

[6] Akyildiz I. F., Vuran M.C. *Wireless Sensor Networks*. John Wiley & Sons Press; 2010.

[7] Wikipedia. *List of Wireless Sensor Nodes* [online]. Available from https://en.wikipedia.org/wiki/List_of_wireless_sensor_nodes [Accessed 26 March 2016]

[8] Hill J., Szewczyk R., Woo A., Hollar S., Culler D., Pister K. 'System architecture directions for networked sensors'. In *Proceedings of the ACM SIGOPS Operating Systems Review*; Cambridge, MA, USA, 12–15 December 2000. pp. 93–104.

[9] Watteyne T., Vilajosana X., Kerkez B., *et al.* 'OpenWSN: a standards-based low-power wireless development environment'. *Transactions on Emerging Telecommunications Technologies*. 2012;23(5):480–493.

[10] Micrium | Real-time Operating Systems. *A Real-time Kernel for Embedded System* [online]. 2016. Available from http://micrium.com [Accessed 26 March 2016]

[11] Han C.C., Kumar R., Shea R., Kohler E., Srivastava M. 'A dynamic operating system for sensor nodes'. In *Proceedings of the 3rd International Conference on Mobile Systems, Applications, and Services (MobiSys 2005)*; Seattle, Washington, USA, 6–8 June, 2005. pp. 163–176.

[12] RTJ Computing Pty. Ltd. *simpleRTJ Technical Brief* [online]. 2009. Available from http://www.rtjcom.com/downfile.php?f=techpdf [Accessed 26 March 2016]

[13] LEGO Mindstorms. *LeJOS, Java for Lego Mindstroms* [online]. 2006. Available from http://www.lejos.org [Accessed 26 March 2016]

[14] Tills Palm Pages. *The NanoVM – Java for the AVR* [online]. 2005. Available from http://www.harbaum.org/till/nanovm/index.shtml [Accessed 26 March 2016]

[15] Oracle Pty. Ltd. *Java Card Platform Specification 2.2.2.* [online]. Available from http://www.oracle.com/technetwork/java/javacard/specs-138637.html [Accessed 26 March 2016]

[16] Brouwers N., Langendoen K., Corke P. 'Darjeeling, a feature-rich VM for the resource poor'. In *Proceedings of Embedded Networked Sensor System*; Berkeley, CA, USA, 4–6 November 2009. pp. 169–182.

[17] Ousterhout J. 'Why threads are a bad idea (for most purposes)'. In *Proceedings of the 1996 USENIX Annual Technical Conference*; San Diego, CA, USA, 22–26 January 1996. Volume 5.

[18] Von Behren R., Condit J., Brewer E. 'Why events are a bad idea (for high-concurrency servers)'. In *Proceedings of the 9th Workshop on Hot Topics in Operating Systems*; Lihue, HI, USA, 18–21 May 2003. pp. 19–24.

[19] Keen A.W., Ishihara T., Maris J.T., Li T., Fodor E.F., Olsson R.A. 'A comparison of concurrent programming and cooperative multithreading'. *Concurrency and Computation: Practice and Experience*. 2003;15(1):27–53.

[20] Chen Y.T., Chien T.C., Chou P.H. 'Enix: a lightweight dynamic operating system for tightly constrained wireless sensor platforms'. In *Proceedings of the 8th ACM Conference on Embedded Networked Sensor Sy*stems; Zurich, Switzerland, 3–5 November, 2010. pp. 183–196.

[21] Klues K., Liang C.J.M., Paek J., *et al.* 'TOSThreads: thread-safe and non-invasive preemption in TinyOS'. In *Proceedings of the ACM Conference on Embedded Networked Sensor Systems*; Berkeley, CA, USA, 4–6 November 2009. pp. 127–140.

[22] Dunkels A., Gronvall B., Voigt T. 'Contiki—a lightweight and flexible operating system for tiny networked sensors'. In *Proceedings of Local Computer Networks*; Tampa, FL, USA, 16–18 November 2004. pp. 455–462.

[23] Zhou H.Y., Hou K.M., De Vaulx C., Zuo D.C. 'A hybrid embedded real-time operating system for wireless sensor networks'. *Journal of Networks*. 2009;4(6):428–435.

[24] Stephen Ferg. *Event-Driven Programming: Introduction, Tutorial, History* [online]. 2006. Available from http://eventdrivenpgm.sourceforge.net [Accessed 26 March 2016]

[25] Liu C.L., Layland J.W. 'Scheduling algorithms for multiprogramming in a hard-real-time environment'. *Journal of the ACM*. 1973;20(1):46–61.

[26] Buttazzo G.C. 'Rate monotonic vs. EDF: judgment day'. *Real-Time Systems*. 2005;29(1):5–26.

[27] Wilson P.R., Johnstone M.S., Neely M., Boles D. 'Dynamic storage allocation: a survey and critical review'. *Memory Management*. Springer Berlin Heidelberg, 1995: pp. 1–116.

[28] Johnstone M.S., Wilson P.R. 'The memory fragmentation problem: solved?'. *ACM SIGPLAN Notices*. 1998;34(3):26–36.

[29] Bhatti S., Carlson J., Dai H., Deng J. 'MANTIS OS: An embedded multi-threaded operating system for wireless micro sensor platforms'. *ACM Kluwer Mobile Networks and Applications*. 2005;10(4):563–579.

[30] Real Time Engineers Ltd. *FreeRTOS – Market Leading RTOS for Embedded Systems with Internet of Things Extensions* [online]. 2010. Available from http://www.freertos.org/ [Accessed 26 March 2016]

[31] Adam Dunkels. *Contiki Managed Memory Allocator* [online]. 2012. Available from https://github.com/adamdunkels/contiki-fork/wiki/Memory-allocation [Accessed 26 March 2016]

[32] Gu L., Stankovic J.A. 't-kernel: providing reliable OS support to wireless sensor networks'. In *Proceedings of the 4th ACM International Conference on Embedded Networked Sensor Systems*; Boulder, Colorado, USA, 31 October–3 November, 2006. pp. 1–14.

[33] Philip Levis. *TinyOS programming* [online]. 2006. Available from http://www.tinyos.net/tinyos-2.x/doc/pdf/tinyos-programming.pdf [Accessed 26 March 2016]

[34] Dunkels A., Schmidt O., Voigt T., Ali M. 'Protothreads: simplifying event-driven programming of memory-constrained embedded systems'. In *Proceedings of the 4th International Conference on Embedded Networked Sensor Systems*; Boulder, Colorado, USA, 31 October–3 November 2006. pp. 29–42.

[35] Wang Q., Zhu Y., Cheng L. 'Reprogramming wireless sensor networks: challenges and approaches'. *IEEE Network Magazine*. 2006;20(3):48–55.

[36] IEEE Standards Association. *IEEE SA-802.15.4-2006-IEEE Standard for Information Technology* [online]. Available from https://standards.ieee.org/findstds/standard/802.15.4-2006.html [Accessed on 26 March 2016]

[37] ZigBee Alliance. *ZigBee 3.0: The Foundation for the Internet of Things* [online]. 2015. Available from http://www.zigbee.org/zigbee-for-developers/zigbee3-0 [Accessed on 26 March 2016]

[38] Montenegro G., Kushalnagar N., Hui J., Culler D. *Transmission of IPv6 packets over IEEE 802.15.4 networks*. Internet proposed standard RFC 4944, 2007.

[39] Panta R.K., Bagchi S., Midkiff S.P. 'Zephyr: efficient incremental reprogramming of sensor nodes using function call indirections and difference computation'. In *Proceedings of USENIX Annual Technical Conference*; San Diego, CA, USA, 14–19 June 2009.

[40] Panta R.K., Bagchi S. 'Hermes: fast and energy efficient incremental code updates for wireless sensor networks'. In *Proceedings of 28th IEEE International Conference INFOCOM*; Rio de Janeiro, Brazil, 19–25 April, 2009. pp. 639–647.

[41] Qiu J., Li D., Shi H., Cui L. 'EasiLIR: lightweight incremental reprogramming for sensor networks'. *International Journal of Distributed Sensor Networks*. 2014; Article ID: 120597, pp. 1–15.

[42] Dong W., Liu Y., Chen C., Bu J., Huang C., Zhao Z. 'R2: incremental reprogramming using relocatable code in networked embedded systems'. *IEEE Transactions on Computers*. 2013;62(9):1837–1849.

[43] Sun J.Z. 'Dissemination protocols for reprogramming wireless sensor networks: a literature survey'. In *Proceedings of IEEE International Conference on Sensor Technologies and Applications*; Venice, Italy, 18–25 July 2010. pp. 151–156.

[44] Stolikj M., Cuijpers P.J.L., Lukkien J.J. 'Energy-aware reprogramming of sensor networks using incremental update and compression'. *Procedia Computer Science*. 2012;10:179–187.

[45] Tsiftes N., Dunkels A., Voigt T. 'Efficient sensor network reprogramming through compression of executable modules'. In *Proceedings of IEEE Conference on Sensor, Mesh and Ad Hoc Communications and Networks*; San Francisco, CA, USA, 16–20 June 2008. pp. 359–367.

[46] Saginbekov S., Jhumka A. 'Towards efficient stabilizing code dissemination in wireless sensor networks'. *Computer Journal*. 2014;57(12): 1790–1816.

[47] Tan H., Ostry D., Zic J., Jha S. 'A confidential and DoS-resistant multi-hop code dissemination protocol for wireless sensor networks'. *Computers and Security*. 2013;32:36–55.

[48] He D., Chen C., Chan S., Bu J. 'DiCode: DoS-resistant and distributed code dissemination in wireless sensor networks'. *IEEE Transactions on Wireless Communications*. 2012;11(5):1946–1956.

[49] Zhang R., Zhang Y. 'LR-Seluge: loss-resilient and secure code dissemination in wireless sensor networks". In *Proceedings of IEEE International Conference on Distributed Computing Systems*; Minneapolis, USA, 20–24 June 2011. pp. 497–506.

[50] Dong W., Liu Y., Wang C., Liu X., Chen C., Bu J. 'Link quality aware code dissemination in wireless sensor networks'. In *Proceedings of IEEE International Conference on Network Protocols*; Vancouver, Canada, 17–20 October 2011. pp. 89–98.

[51] Raghunathan V., Ganeriwal S., Srivastava M. 'Emerging techniques for long lived wireless sensor networks'. *IEEE Communications Magazine*. 2006;44(4):108–114.

[52] Simon G., Maróti M., Lédeczi Á., *et al.* 'Sensor network-based countersniper system'. In *Proceedings of the International Conference on Embedded Networked Sensor Systems*; Baltimore, Maryland, USA, 3–5 November 2004. pp. 1–12.

[53] Kanagal B., Deshpande A. 'Online filtering, smoothing and probabilistic modeling of streaming data'. In *Proceedings of the 24th International Conference on Data Engineering*; Cancún, México, April 2008. pp. 1160–1169.

[54] Jain A., Chang E.Y., Wang Y.F. 'Adaptive stream resource management using Kalman filters'. In *Proceedings of the ACM International Conference on Management of Data*; Paris France, 13–18 June, 2004. pp. 11–22.

[55] Pottie G., Kaiser W. 'Wireless integrated network sensors'. *Communication of ACM*. 2000;43(5):51–58.

[56] Caione C., Brunelli D., Benini L. 'Distributed compressive sampling for lifetime optimization in dense wireless sensor networks'. *IEEE Transactions on Industrial Informatics*. 2012;8(1):30–40.

[57] Srisooksai T., Keamarungsi K., Lamsrichan P., Araki K. 'Practical data compression in wireless sensor networks: a survey'. *Journal of Network and Computer Applications*. 2012;35(1):37–59.

[58] Maraiya K., Kant K., Gupta N. 'Wireless sensor network: a review on data aggregation'. *International Journal of Scientific and Engineering Research*. 2011;2(4):1–6.

[59] Xiang L., Luo J., Vasilakos A. 'Compressed data aggregation for energy efficient wireless sensor networks'. In *Proceedings of IEEE Conference on Sensor, Mesh and Ad hoc Communications and Networks*; Salt Lake, Utah, USA, 27–30 June, 2011. pp. 46–54.

[60] Aziz A.A., Sekercioglu Y.A., Fitzpatrick P., Ivanovich M. 'A survey on distributed topology control techniques for extending the lifetime of battery powered wireless sensor networks'. *Communications Surveys and Tutorials*. 2013;15(1):121–144.

[61] Li M., Li Z., Vasilakos A.V. 'A survey on topology control in wireless sensor networks: taxonomy, comparative study, and open issues'. *Proceedings of the IEEE*. 2013;101(12):2538–2557.

[62] Üster H., Lin H. 'Integrated topology control and routing in wireless sensor networks for prolonged network lifetime'. *Ad Hoc Networks*. 2011;9(5):835–851.

[63] Warrier A., Park S., Mina J., Rheea I. 'How much energy saving does topology control offer for wireless sensor networks? – a practical study'. *ACM Computer Communications*. 2007;30(14):2867–2879.

[64] Ganesan D., Cerpa A., Ye W., Yu Y., Zhao J., Estrin D. 'Networking issues in wireless sensor networks'. *Journal of Parallel and Distributed Computing*. 2004;64(7):799–814.

[65] Saleh A.M.S, Ali B.M., Rasid M.F.A, Ismail A. 'A survey on energy awareness mechanisms in routing protocols for wireless sensor networks using optimization methods'. *Transactions on Emerging Telecommunications Technologies*. 2014;25(12):1184–1207.

[66] Gilani M.H.S., Sarrafi I., Abbaspour M. 'An adaptive CSMA/TDMA hybrid MAC for energy and throughput improvement of wireless sensor networks'. *Ad Hoc Networks*. 2013;11(4):1297–1304.

[67] Rajendran V., Obracza K., Garcia-Luna Aceves J.J. 'Energy-efficient, collision-free medium access control for wireless sensor networks'. In *Proceedings of International Conference on Embedded Networked Sensor Systems*; Los Angeles, USA, 5–7 November 2003. pp. 63–78.

[68] Ye W., Heidemann J., Estrin D. 'Medium access control with coordinated adaptive sleeping for wireless sensor networks'. *IEEE/ACM Transactions on Networking*. 2004;12(3):493–506.

[69] Zareei M., Taghizadeh A., Budiarto R., Wan T.C. 'EMS-MAC: energy efficient contention-based medium access control protocol for mobile sensor networks'. *Computer Journal*. 2011.

[70] Pantazis N., Nikolidakis S.A., Vergados D.D. 'Energy-efficient routing protocols in wireless sensor networks: a survey'. *IEEE Communications Surveys and Tutorials*. 2013;15(2):551–591.

[71] Eswaran A., Rowe A., Rajkumar R. 'Nano-RK: an energy-aware resource-centric RTOS for sensor networks'. In *26th IEEE International on Real-Time Systems Symposium*; Miami, FL, USA, 5–8 December, 2005. pp. 10–19.

[72] Xia F., Hao R., Li J., Xiong N., Yang L.T., Zhang Y. 'Adaptive GTS allocation in IEEE 802.15. 4 for real-time wireless sensor networks'. *Journal of Systems Architecture*. 2013;59(10):1231–1242.

[73] Azeem M., Khan M.I., Khan S.U., Gansterer W. 'Efficient scheduling of sporadic tasks for real-time wireless sensor networks'. *IET Wireless Sensor Systems*. 2014;5(1):1–10.

[74] Mouradian A., Augé-Blum I., Valois F. 'RTXP: a localized real-time MAC-routing protocol for wireless sensor networks'. *Computer Networks*. 2014;67:43–59.

[75] Aissani M., Bouznad S., Fareb A., Laidoui M.A. 'EA-SPEED: energy-aware real-time routing protocol for wireless sensor networks'. *International Journal of Information and Communication Technology*. 2013;5(1):22–44.

[76] B. Shah, K.I. Kim, "A new real-time and guaranteed lifetime protocol in wireless sensor networks". *International Journal of Distributed Sensor Networks*. 2014.

[77] Szalapski T., Madria, S. 'On compressing data in wireless sensor networks for energy efficiency and real time delivery'. *Distributed and Parallel Databases*. 2013;31(2):151–182.

[78] Chouikhi S., El Korbi I., Ghamri-Doudane Y., Saidane L.A. 'A survey on fault tolerance in small and large scale wireless sensor networks'. *Computer Communications*. 2015;69:22–37.

[79] Munir A., Antoon J., Gordon-Ross A. 'Modeling and analysis of fault detection and fault tolerance in wireless sensor networks'. *ACM Transactions on Embedded Computing Systems*. 2015;14(1):1–49.

[80] Tolle G., Polastre J., Szewczyk R., *et al.* 'A macroscope in the redwoods'. In *Proceedings of the 3rd International Conference on Embedded Networked Sensor Systems*; San Diego, USA, 2–4 November, 2005. pp. 51–63.

[81] De Souza L.M.S., Vogt H., Beigl M. 'A survey on fault tolerance in wireless sensor networks'. *SAP Research*. 2007.

[82] Chanak P., Banerjee I. 'Energy efficient fault-tolerant multipath routing scheme for wireless sensor networks'. *Journal of China Universities of Posts and Telecommunications*. 2013;20(6):42–61.

[83] Younis M., Senturk I.F., Akkaya K., Lee S., Senel F. 'Topology management techniques for tolerating node failures in wireless sensor networks: a survey'. *Computer Networks*. 2014;58:254–283.

[84] Cha H., Choi S., Jung I., *et al.* 'RETOS: resilient, expandable, and threaded operating system for wireless sensor networks'. In *Proceedings of International Symposium on Information Processing in Sensor Networks*; Cambridge, MA, USA, 25–27 April 2007. pp. 148–157.

[85] Cao Q., Abdelzaher T., Stankovic J., He T. 'The LiteOS operating system: towards UNIX-like abstractions for wireless sensor networks'. In *International Conference on Information Processing in Sensor Networks*; St. Louis, Missouri, USA, 22–24 April, 2008. pp. 233–244.

Chapter 3

Wireless sensor network operating system: concept, new design, and implementation

Xing Liu[1,5]*, Haiying Zhou*[2]*, Shen Lin*[3]*, Shengwu Xiong*[1]*, Jian Li*[4] *and Kun-Mean Hou*[5]

Abstract

Memory optimization, real-time scheduling, energy conservation, reprogramming, context awareness, and fault tolerance are the critical research challenges in the wireless sensor network (WSN). To address these challenges, a new WSN operating system (OS) LiveOS is designed and implemented in this chapter. Compared with the other WSN OSes, LiveOS has two typical features. On the one hand, the new OS design concepts such as the hybrid scheduling, the shared-stack scheduling, the reactive-defragmentation allocation, and the pre-linked native-code middleware are implemented. By doing this, the memory cost of the real-time WSN OS can be decreased. Moreover, the reprogramming performance of the WSN nodes can be improved. On the other hand, the new research approach, which addresses the WSN challenges by combining both the software technique and the multi-core hardware technique, is applied in LiveOS. By means of the multi-core hardware infrastructure, the lifetime of the LiveOS node can be prolonged. Moreover, the context-aware ability, the real-time performance, and the fault-tolerant capability of the WSN nodes can be improved. With the implementation of the above concepts, LiveOS becomes the WSN OS which can be applied on the resource-constrained WSN nodes and can be used to execute the real-time WSN applications with high reliability.

[1]Hubei Key Laboratory of Transportation Internet of Things, Wuhan University of Technology, 430070 Wuhan, China
[2]School of Electrical and Information Engineering, HuBei University of Automotive Technology, 442002 Shiyan, HuBei, China
[3]Key Laboratory of Automobile Parts Testing, General Administration of Quality Supervision, Inspection and Quarantine of P.R. China, 441003 XiangYang, HuBei, China
[4]School of Computer Science, Harbin Institute of Technology, 150001 Harbin, China
[5]LIMOS Laboratory, UMR 6158 CNRS, Blaise Pascal University, Les Cézeaux, BP 10125, 63173 Clermont-Ferrand, France

3.1 Introduction

WSN is composed of a set of nodes which can monitor the environmental conditions and pass the collected data cooperatively to a main location [1]. Currently, the WSN technology has been applied in widespread application domains such as the precision agriculture, the disaster management, the intelligent transport system, the industrial control, and the medical care [2,3].

Due to its typical features, many research challenges exist in the WSN. First, the WSN nodes need to be low cost and small size so as to keep in with the strong market competitive strength. With the cost and size limitation, most WSN nodes are constrained in the memory resource [1], e.g., the MicaZ node has only 4 KB RAM. Thus, the memory optimization becomes significant for the WSN nodes. Second, the WSN nodes are commonly powered by the energy-limited batteries and are difficult to be recharged after being deployed. Therefore, the energy conservation becomes important to the WSN nodes [4]. Third, the WSN platforms are diverse. Different hardware and software platforms have been developed [5]. Thus, the user application programming is difficult. To simplify the programming complexity, the research of a mechanism which can abstract the low-level system details becomes essential. Fourth, with the proliferation of the WSN technique, the WSN applications become more and more comprehensive. Currently, many WSN applications start to merge with the wireless multimedia sensor network (WMSN) applications. Since the WMSN applications have the run-time contexts different from the WSN applications, the research of the new WSN OS which can be context aware to the comprehensive application contexts become important. Finally, the WSN nodes are prone to be deployed in the harsh environments such as the mine and agriculture field, and they are difficult to be maintained after the deployment. Therefore, the remote reprogramming and the fault-tolerant ability become significant to the WSN nodes.

To prompt the proliferation of the WSN and address the challenges above, the design and implementation of an outstanding WSN OS becomes critical. With an outstanding WSN OS, not only the constrained WSN platform resources can be managed efficiently, but also the complicated WSN application development can be simplified soundly. Due to this reason, a serial of WSN OSes have been developed in the past, including the TinyOS [6], the Contiki [7], the SOS [8], the MantisOS [9], the openWSN [10], the Enix [11], and the LiteOS [12]. These OSes have different features and prompted the proliferation of the WSN technology. Yet, some improvement works still needs to be achieved.

In this chapter, a new WSN OS termed LiveOS is designed and implemented. Different from the other WSN OSes, LiveOS has two typical features. On the one hand, the new design concepts, such as the hybrid scheduling, the shared-stack scheduling, the reactive-defragmentation allocation and the pre-linked native-code middleware are implemented. With these mechanisms, the memory cost of LiveOS can be optimized significantly. Moreover, a user-friendly application development environment can be provided to the users. On the other hand, the new research approach which addresses the WSN challenges by combining both the software technique and the multi-core hardware technique is applied in LiveOS. By means of the multi-core hardware infrastructure, the energy cost, the real-time performance, the context-aware ability, the fault-tolerant performance, and the debugging functionality of the WSN OSes can be optimized soundly. With the above features, LiveOS becomes the WSN OS which is real time, memory efficient, energy efficient, context aware, fault tolerant, and user friendly. In addition, it can run on most WSN nodes to serve the different WSN applications effectively.

The structure of this chapter is organized as follows: In section 3.2, the hybrid scheduling and the shared-stack scheduling mechanisms of LiveOS are presented. With these mechanisms, the memory consumption of LiveOS can be decreased significantly whereas the real-time performance of LiveOS can keep well. In section 3.3, the reactive-defragmentation memory allocator of LiveOS is introduced. With this allocator, the memory fragments during the allocation time can be defragmented, yet the allocation overhead still keeps not high. In section 3.4, a resource-efficient middleware LiMid, which decouples the application code from the low-level system code, is implemented and embedded inside LiveOS. With the LiMid, not only the LiveOS application programming complexity can be simplified, but also the application reprogramming performance can be optimized. In section 3.5, the multi-core task assignment mechanism which can be used to optimize the energy cost of the WSN tasks is realized. By doing this, the lifetime of the WSN nodes can be prolonged. In section 3.6, the approach of using the multi-core task assignment to improve the OS real-time performance in the LiveOS is introduced. In section 3.7, the LiveOS multi-core context-aware platform, with which the node can self-adjust to become aware to the different application contexts, is investigated. In section 3.8, the works of using the multi-core technique to validate the run-time status of the microcontroller and recover the run-time faults of the WSN nodes are investigated. By this multi-core checking and recovery techniques, the availability of the WSN nodes can be improved significantly. In section 3.9, the LiveOS multi-core debugging approach is presented. With this debugging approach, the online debugging on the high memory-constrained WSN node can be realized efficiently. In section 3.10, a discussion about the LiveOS design concepts is presented. Finally in section 3.11, the conclusion and the ongoing works of LiveOS are concluded.

3.2 LiveOS memory-efficient real-time scheduling

To achieve the real-time scheduling, the multithreaded preemption is needed. Yet, when a thread is preempted, its run-time context needs to be stored. As a result, each thread in the multithreaded OS needs to be allocated an independent stack, and the memory cost of the multithreaded OS becomes high [13,14]. To run the multithreaded OS on the tightly resource-constrained WSN nodes, the memory optimization to the thread stacks is essential, and this has currently been realized in LiveOS by using the hybrid scheduling and the shared-stack multithreading. With the hybrid scheduling, the number of the stacks in the multithreaded OS can be decreased With the shared-stack multithreading, the average size of the stacks in the multithreaded OS can be reduced. By the above means, the memory cost of the multithreaded OS can be optimized significantly.

3.2.1 Hybrid scheduling

The objective of LiveOS hybrid scheduling is to implement a hybrid scheduler which targets to combine the advantage of the multithreaded OS good real-time performance and the event-driven OS low memory cost. With this hybrid scheduler, the stack number of the multithreaded OS can be optimized. Consequently, the real-time multithreaded OS can run even on the high memory-constrained WSN nodes.

The concept of LiveOS hybrid scheduling is to strip the redundant thread stacks which can exist in the traditional multithreaded OSes. In the multithreaded OS, each thread runs in its own independent stack and this stack will be used to save the thread's context when the thread is preempted. Yet, not all threads need the preemption functionality during the run-time. In case that the threads have loose requirement to the responsive time, they can be executed one by one without preemption. Due to this reason, the tasks in the LiveOS are classified into two kinds: the real-time tasks and the non-real-time tasks. The real-time tasks need to run to completion before the deadline, thus they are scheduled by the preemptive multithreaded scheduler and each task takes up an independent stack. The non-real-time tasks do not have the strict restriction to the responsive time, thus they can be executed one by one without preemption and be scheduled by the event-driven scheduler [15]. Since all non-real-time tasks which are managed by the event-driven scheduler, run within a shared stack in LiveOS, the stack number of the multithreaded LiveOS can be decreased. As a result, the stack memory cost of LiveOS can be optimized. In Figure 3.1, the elementary diagram of the LiveOS hybrid scheduling is depicted. It is assumed that there are six non-real-time tasks and three real-time tasks in the system. If the traditional multithreaded OS is used to schedule these tasks, nine threads along with nine stacks need to be allocated. Yet, after the LiveOS hybrid scheduling is applied, all the non-real-time tasks run one by one within a shared stack. As a result, only four stacks need to be created. Since the thread stack is not small in the data memory size, the reduction of the stack number can optimize the stack memory cost significantly.

In the hybrid LiveOS, two kinds of schedulers exist. To simplify the implementation of this hybrid scheduler, the whole LiveOS event-driven scheduling system is also implemented as a thread, named *nonRT_thread*. By doing this, the implementation of the LiveOS hybrid scheduling system can be implemented as that of the general multithreaded system, and the implementation complexity can be eased, depicted in Figure 3.2. Currently, the static-scheme fixed-priority scheduling algorithm RMS

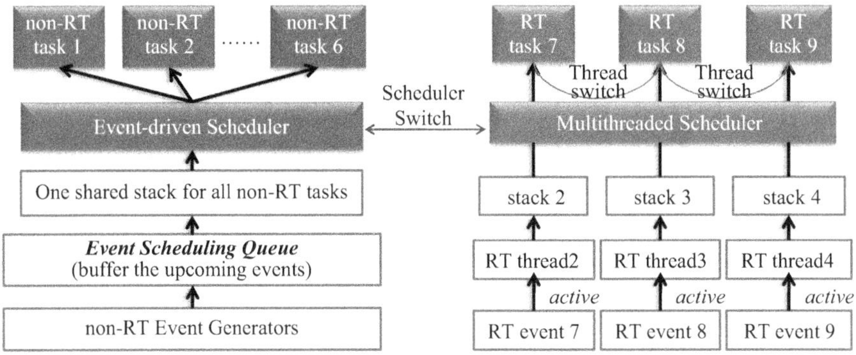

Figure 3.1 Elementary diagram of the LiveOS hybrid scheduling. If six non-real-time tasks and three real-time tasks exist in LiveOS, only four stacks need to be allocated. Yet, in the traditional multithreaded OS, nine threads along with nine stacks need to be created

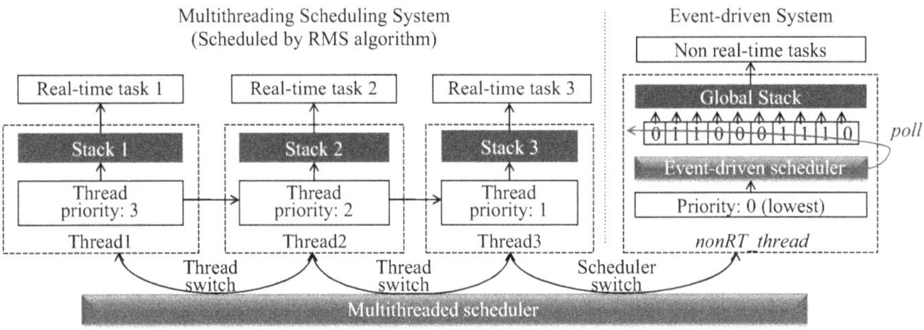

Figure 3.2 Implementation of the LiveOS hybrid scheduling. The event-driven scheduling system is also implemented as a thread. By doing this, the switch between these two schedulers can be operated as the switch from one thread to the other

[16] is applied in the hybrid LiveOS. The non-real-time *nonRT_thread* has the lowest priority. It can be preempted anytime by the real-time threads, and it can be executed only when all the real-time tasks are inactive.

3.2.2 Shared-stack multithreading

Besides the optimization to the stack number, the optimization to the average stack size is also significant to reduce the OS stack memory cost. Currently, this has been realized in the LiveOS by implementing the shared-stack multithreading mechanism.

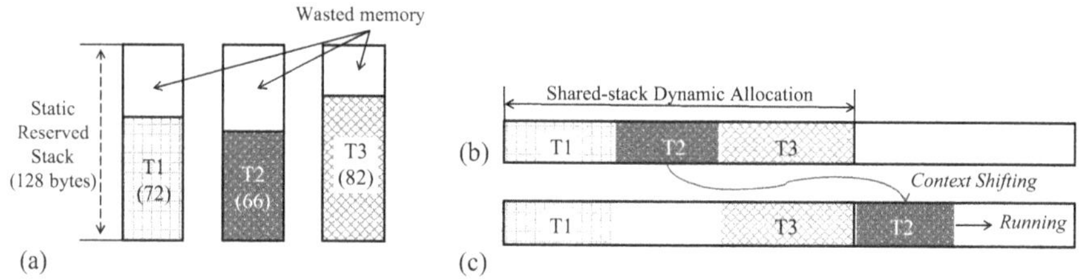

Figure 3.3 (a) Stack memory will be wasted in the traditional multithreaded OS due to the static pre-reservation mechanism. (b) Stacks are allocated dynamically within the shared stack in LiveOS. (c) Context shifting technique is applied to avoid the corruption to the context data

In traditional multithreaded system, each thread is allocated an independent stack, and the stack memory is pre-reserved statically. Since the stacks are allocated statically, the stack size needs to be large enough to meet the worst-case scenario. As a result, the stack memory cannot be avoided to be wasted (Figure 3.3). One way to solve this problem is to run all the threads within a shared stack, and allocate the memory space of each stack dynamically as it is required. By doing this, the stack memory space which can be wasted in the traditional multithreaded WSN OS can be avoided.

However, the threads' stacks are closely adjacent to each other after the shared-stack multithreading mechanism is used, and this will cause the stack data of one thread to be corrupted when the other threads resume its execution, e.g., in Figure 3.3 the stack data of $T3$ will be corrupted if the $T2$ resumes its execution. To solve this problem, the context shifting technique is applied in the shared-stack LiveOS. Once an inactive thread is ready to be executed, its context will be shifted to the free memory space (Figure 3.3). By doing this means, the context corruption problem described above can be avoided.

After the contexts shifting is carried out for a period of time, the left free stack space cannot be sufficient for the new thread to run. In this case, the memory defragmentation to the shared stack needs to be performed. In LiveOS, the defragmentation to the shared stack is operated the same as that to the heap space (Section 3.3).

3.2.3 Performance evaluation

In this section, the scheduling performance of the LiveOS hybrid scheduler and the LiveOS shared-stack scheduler is evaluated by comparing with the TinyOS, the Contiki, the SOS, the MantisOS, and the TOSThread [17] from the perspectives of code memory cost, data memory cost, and scheduling efficiency. The evaluation is performed on the iLive node (Figure 3.4), and the AVR studio is used as the compiler.

The code memory cost of the different schedulers is shown in Figure 3.5. SOS is a modular-architecture OS in which the module management mechanism needs to be

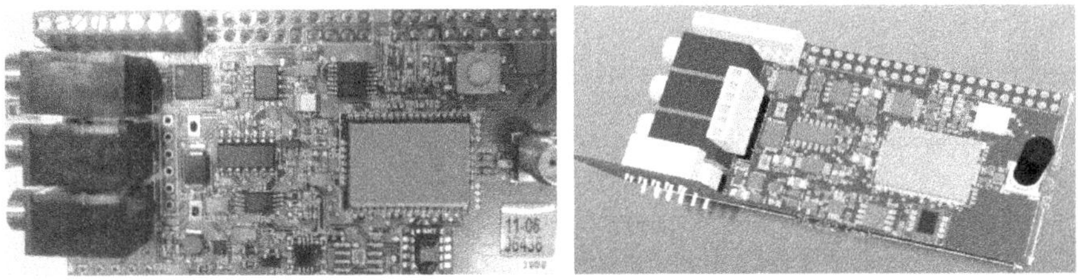

Figure 3.4 Prototype board of the iLive node. ILive node is equipped with the 8-bit AVR ATmega1281 microcontroller

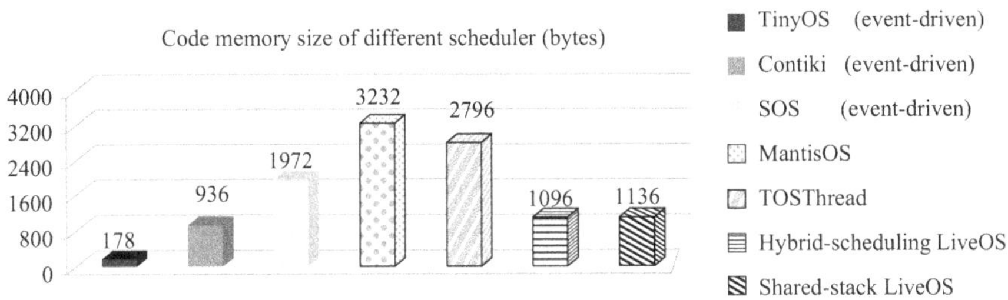

Figure 3.5 Code memory cost of the schedulers in different WSN OSes

implemented. As a result, the code memory cost of SOS is higher than that of Contiki and TinyOS. MantisOS has the code memory cost higher than that of LiveOS. This is because MantisOS is a traditional multithreaded OS. In MantisOS, all the tasks need to be executed by threads. As a result, the thread number is large and the complicated multilevel queue scheduling mechanism is implemented. TOSThread also has the code memory cost higher than that of LiveOS, and this is because a flexible application and system boundary which can simplify the application development in TinyOS is implemented in the TOSThread.

The data memory cost in the event-driven OSes (TinyOS, Contiki, and SOS) and the multithreaded OSes (TOSThread, MantisOS, and LiveOS) can be denoted respectively as follows:

$$M_{EVENT} = S_{DATA/BSS} + S_{PCB} \times N_{PCB} + S_{EVT} \times L_{SQ} + S_{STK}$$

$$\begin{aligned} M_{THREAD} &= S_{DATA/BSS} + S(TCB) + S(STK) \\ &= S_{DATA/BSS} + S_{TCB} \times N_{TCB} + S_{STK} \times N_{STK} \end{aligned}$$

in which $S_{DATA/BSS}$ represents the size of the DATA/BSS sections; S_{PCB}, S_{TCB}, S_{EVT} and S_{STK} represent, respectively, the process control block (PCB) size, the thread control block (TCB) size, the event structure size, and the thread stack size; N_{PCB}, N_{TCB}

L_{SQ}, and N_{STK} represent, respectively, the PCB number, the TCB number, the event scheduling queue (SQ) length, and the thread stack number. It is assumed that the N_{PCB} and L_{SQ} in the event-driven OSes TinyOS, Contiki, and SOS are, respectively, 12 and 10; the N_{TCB}, S_{STK}, and N_{STK} in the traditional multithreaded OSes TOSThread and MantisOS are, respectively, 12, 128, and 10; the N_{TCB}, S_{STK}, and N_{STK} in the hybrid-scheduling LiveOS are, respectively, 8, 128, and 6 (thread number of the hybrid LiveOS can be smaller than that of the traditional multithreaded OS); the N_{TCB}, S_{STK}, and N_{STK} in the shared-stack multithreading LiveOS are, respectively, 12, 80, and 10 (average stack size of the shared-stack multithreading LiveOS can be smaller than that of the traditional multithreaded OS). Then, the data memory cost of different schedulers can be calculated, depicted in Figure 3.6.

The scheduling efficiency of different schedulers can be evaluated by the scheduling clock cycles. In the event-driven systems, the scheduling processes include the operation of extracting the events and the dispatch of the events to the related tasks [15]. In the multithreaded OSes, the scheduling processes include the operation of switching the contexts and the operation of selecting the next thread to be scheduled. In Figure 3.7, the scheduling clock cycles of different schedulers are shown. In the shared-stack multithreading LiveOS, the extra scheduling overhead caused by the context shifting exists, thus the scheduling overhead becomes higher.

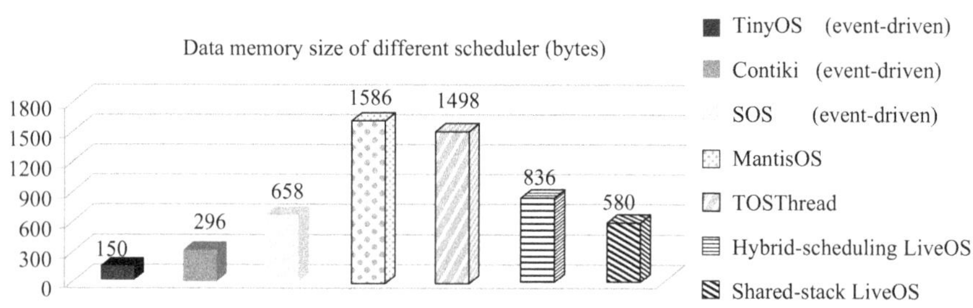

Figure 3.6 Data memory cost of the schedulers in different WSN OSes

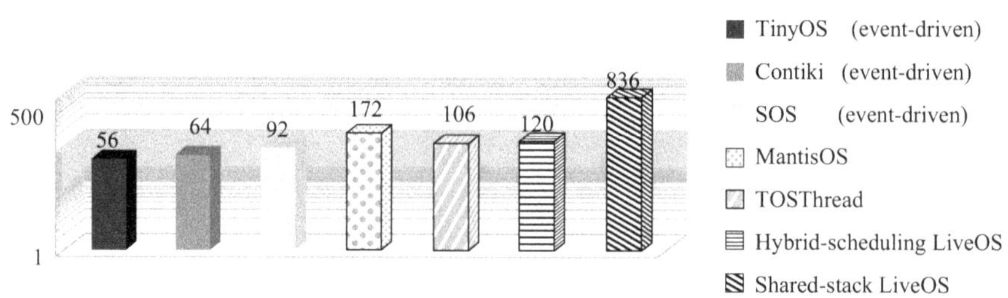

Figure 3.7 Scheduling clock cycles of the different schedulers

3.2.4 Discussion

From the evaluations above, it is shown that the data memory cost of the multithreaded scheduler is commonly higher than that of the event-driven scheduler, and this is because more stacks need to be allocated in the multithreaded system.

With the hybrid scheduling mechanism, the stack number of LiveOS can be smaller. As a result, the data memory cost of LiveOS can be 47.3 per cent lower than that of traditional multithreaded MantisOS. With the shared-stack scheduling mechanism, the average stack size of LiveOS can be decreased. Consequently, the data memory cost of LiveOS can be 63.4 per cent lower than that of traditional MantisOS.

In the shared-stack multithreaded LiveOS, the stack memory cost can be optimized significantly. Yet, the scheduling overhead will increase due to the context shifting operation. Therefore, the optimization to the scheduling overhead of the shared-stack LiveOS becomes essential. Fortunately, this optimization can be realized by combining both the hybrid scheduling and the shared-stack scheduling. With the hybrid scheduling, the thread switch will be performed only among the real-time tasks, and the thread switch frequency will be reduced greatly. Consequently, the entire scheduling overhead of the shared-stack LiveOS can be optimized soundly.

3.3 LiveOS reactive-defragmentation dynamic memory allocation

Since the RAM resources of the WSN nodes are commonly constrained, the dynamic memory allocation is popularly implemented in the WSN OSes. In the Contiki and the MantisOS, the simple segregated storage allocation and the sequential fit allocation [18] are performed respectively. With these dynamic allocation mechanisms, the memory utilization efficiency of the RAM resources can be improved. Yet, the memory fragments cannot be defragmented by these mechanisms and this lowers the utilization rate of the memory resource. To defragment the fragments, a new allocation mechanism termed *mmem* is implemented in Contiki [19]. With the *mmem*, the fragments can be defragmented proactively. Once a block is released, the allocated blocks adjacent to it will be coalesced. By doing this, all the free blocks will be continuous in the memory address and no fragments will exist.

Although the Contiki *mmem* can perform the defragmentation, its run-time overhead is too high due to the frequent data shifting operations. Therefore, this mechanism is not ideal for the high resource-constrained WSN platform, and the implementation of a new allocation mechanism which can achieve the defragmentation and yet keeps lower allocation overhead becomes essential.

In this section, a new LiveOS memory allocation scheme *lmem*, which is realized by extending the MantisOS sequential fit allocator with the defragmentation functionality, is presented. Compared to the Contiki *mmem*, LiveOS *lmem* is different in that it does not defragment the fragments in the proactive way but in the reactive way.

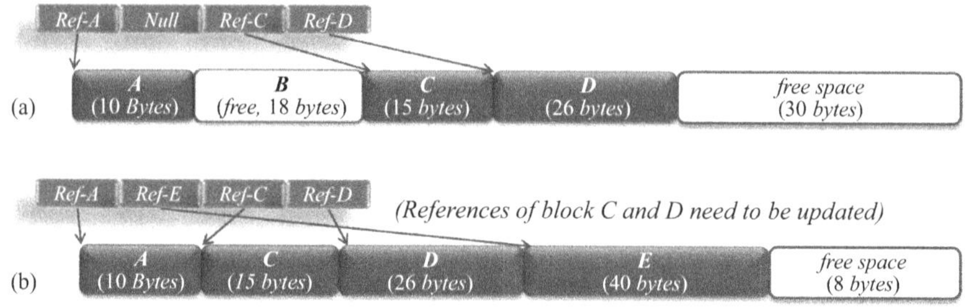

Figure 3.8 LiveOS reactive defragmentation mechanism. Only after the first time allocation is failed, the defragmentation will be performed

By this reactive defragmentation mechanism, the allocation overhead of LiveOS can be optimized.

3.3.1 LiveOS reactive-defragmentation allocation mechanism

In LiveOS, each time a block is released, the coalescence to its adjacent blocks will not be performed immediately. Instead, the coalescence is performed only when the left free blocks are not large enough for the new allocation, e.g., in Figure 3.8, if a 20-byte block needs to be allocated, it can be performed successfully without any defragmentation. Yet, if a 40-byte block needs to be allocated, the allocation will fail since there is no continuous 40-byte free memory. In this case, the coalescence to the blocks *A*, *C* and *D* will be performed. By this means, the defragmentation data shifting frequency can be decreased, and the defragmentation overhead can be optimized.

3.3.2 Performance evaluation

In this section, the allocation performance of the Contiki simple segregated storage, the SOS simple segregated storage, the MantisOS sequential fit, the Contiki *mmem* allocator, and the LiveOS *lmem* allocator is evaluated. The evaluation is performed from the perspectives of the code memory size, the memory utilization efficiency and the allocation overhead.

The code memory sizes of different allocators are shown in Figure 3.9. The LiveOS *lmem* allocator has the code memory size larger than that of MantisOS sequential fit, and this is because the memory defragmentation functionality is implemented extendedly in the LiveOS.

With the results in Figure 3.9, it shows that the memory cost of the dynamic allocators are not so high that the dynamic allocation is feasible to be applied on the resource-constrained WSN nodes. This result has also been recognized by the author of the article [18]. In this article, the author insists that the overhead of the dynamic allocators has been simply overestimated in the past.

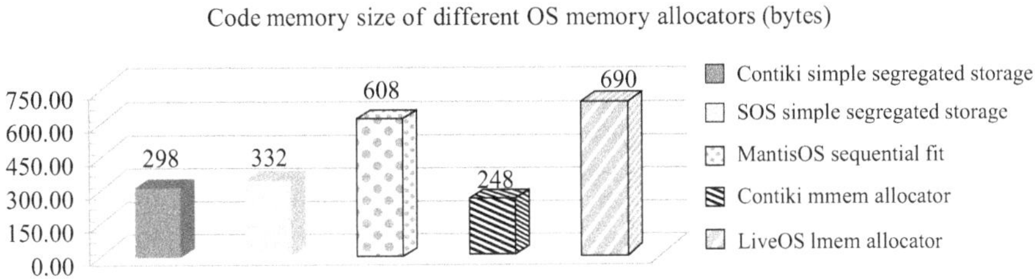

Figure 3.9 Code memory cost of the different memory allocators

To evaluate the memory utilization efficiency and the allocation overhead of different allocators, it is assumed that there exists an allocation scenario as follows:

- The heap memory size is 900 bytes.
- The new blocks to be allocated are sequentially 110, 50, 165, 130, 195, 125, 80, 150, 155, 185, 120, 45, 85, 35, 95, 65, and 105 bytes.
- The allocation is performed every 5 s, and each allocated block will be released 6 s later after being allocated.
- The heap space is divided into three partitions in the Contiki simple segregated storage: 200-byte partition (two items), 125-byte partition (three items), and 50-byte partition (two items).

Based on the above scenario, it is experimented that the allocation in the Contiki simple segregated storage system and the MantisOS sequential fit system will fail respectively at the fifth allocation and the tenth allocation, whereas all the allocations will be performed successfully in the Contiki *mmem* and the LiveOS *lmem* systems. This result proves that the Contiki *mmem* and the LiveOS *lmem* have the memory utilization efficiency higher than the others, and this is because the memory can be fragmented in these two mechanisms.

As for the allocation overhead, it is calculated from the above scenario that the data shifting sizes of the Contiki *mmem* and the LiveOS *lmem* during the defragmentation process are, respectively, 4465 bytes and 365 bytes. From this result, it is shown that the allocation overhead of the LiveOS *lmem* is much lower than that of the Contiki *mmem*, and this is because the reactive defragmentation mechanism, rather than the proactive defragmentation, is realized in the LiveOS.

3.3.3 Discussion

Different allocation mechanisms have different advantages and drawbacks. The simple segregated storage has short and deterministic allocation time, yet the memory utilization efficiency is not high since both external and internal fragments exist. The Contiki *mmem* system has high memory utilization efficiency as the external fragments can be defragmented. Moreover, the allocation time is also deterministic. However, its allocation overhead is high due to the frequent memory coalescence.

The LiveOS *lmem* system has sound memory utilization efficiency since the defragmentation is supported. Moreover, the allocation overhead keeps low as the fragments are defragmented reactively. Yet, its allocation time cannot be deterministic. As for the selection of an appropriate allocation mechanism, it depends on the case of the application requirements and the platform memory resources.

3.4 LiveOS middleware for user-friendly application development environment

Middleware is the intermediate software which lies between the application space and the low-level system space. With the middleware, the application code can be decoupled from the underlying system code. Moreover, a set of abstract programming interfaces can be provided from the system space to the user application space. By so doing, the WSN application programming can be simplified since the WSN users only need to focus on the application development without the necessity of understanding the low-level details. In addition, the WSN reprogramming performance can be improved as only the application binary rather than the monolithic software binary needs to be updated during the reprogramming process. Commonly, the embedded Java VM (EJVM) mechanism [20–23] and the dynamic linking mechanism (DLM) [7,17] have been used in the WSN to decouple the application code from the system code (Figure 3.10). These mechanisms are effective. However, some drawbacks still exist. First, the extra data information, which will be needed either for the bytecode interpretation or the dynamic reference resolving, should be contained in the application code of these mechanisms. As a result, the application code sizes of these mechanisms are not small, and this will increase the energy cost during the reprogramming process. Second, the bytecode interpreter or the dynamic linker need to be implemented in these mechanisms. Consequently, more memory will be cost by the implementation of these mechanisms. Third, the Java bytecode needs to be executed in the interpreted way. As a result, more energy will be cost during the application execution process. Due to the above reasons, the EJVM and the DLM

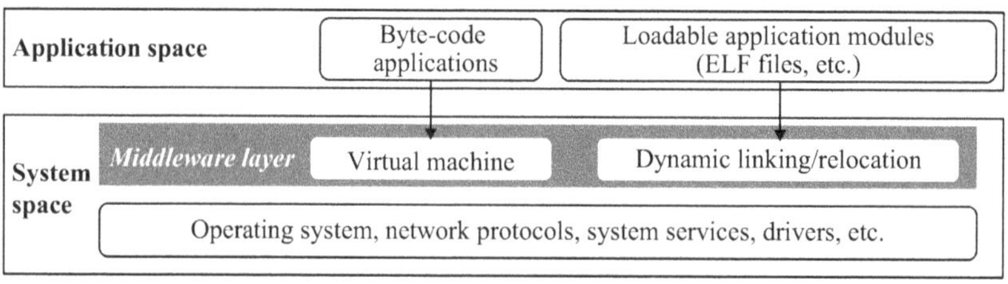

Figure 3.10 Different mechanisms which can decouple the application code from the system code in the WSN

mechanisms are not ideal for the high memory and energy constrained WSN nodes. And the implementation of a new middleware mechanism, which can decouple the application code from the system code and meanwhile keeps low memory and energy cost, becomes essential. In this section, a LiveOS middleware named LiMid will be discussed and evaluated.

3.4.1 LiveOS memory-efficient and energy-efficient middleware LiMid

To develop a middleware which can run even on the tightly resource-constrained WSN nodes, a new software LiMid is developed and embedded inside LiveOS. Similar to the EJVM and the DLM mechanisms, LiMid bridges the interaction between the application space and the system space. Yet, it avoids the drawbacks of both the EJVM and the DLM mechanisms, and keeps low memory and energy cost.

In Table 3.1, the design concept of LiMid is depicted. Different from the EJVM which uses the interpreted bytecode or the DLM which uses the dynamic loadable module, LiMid uses the pre-linked machine-code mechanism to generate the application binary. During the development process, the application code is built independently from the system code. Then, all the system-call functions in the application binary will be re-linked to the corresponded service functions in the system code. With the above means, the application reprogramming code size will become smaller in the LiveOS. Moreover, the code run-time memory and energy cost can keep lower.

However, the pre-linked machine code is less flexible if compared to the bytecode or the dynamic loadable code. To solve this problem, an intermediate function jump table, which locates in a fixed address in the system space, is applied in LiMid (Figure 3.11). With this table, the application system-call functions will not link hardly to the corresponded system service functions, but link indirectly to a given address within the jump table. By doing this, the change to the system code will no more cause the application code to be invalid, and the flexibility of the application binary becomes better.

With the pre-linking mechanism, the application code can call to the system code. Yet, the callback from the system code to the application cannot be achieved. To solve this problem, a callback registration mechanism is implemented in LiveOS. With this mechanism, any registered application function can be called back from the system space (Figure 3.11). By achieving this, the application development process can become easier to the WSN users.

Not only the system callback functionality is realized in LiMid, the multitasking application programming can also be supported by LiMid, and this is realized by the multitasking registration mechanism. In the application space, a set of independent tasks can be defined. Once these tasks are registered by calling the task registration interface, their addresses will be passed to the OS scheduling center, and then they can be scheduled and executed at the time.

Table 3.1 Design concepts of the LiveOS middleware LiMid

Design topics	EJVM mechanism (simpleRTJ)	DLM mechanism (Contiki)	LiMid mechanism				
			Design concept	Why LiMid is designed in this way?			
				Memory efficient	Energy efficient	Small size	Required by WSN
APP file format	Java bytecode	ELF module	Machine code		√	√	
APP executable format		Machine code				N/A	
APP language	Java	C	C				
Call from the APP to the system	By interpretation	By dynamic linking	By pre-linking	√	√		
Callback from the system to the APP	Support	Not support	Support, by registration mechanism				√
APP multitasking							√
Exception handling	Support	Not support				N/A	

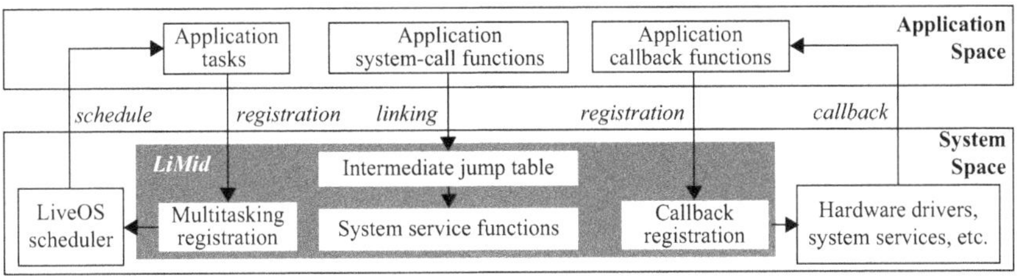

Figure 3.11 Elementary diagram of the LiveOS middleware LiMid

3.4.2 Performance evaluation

In this section, the performance of the LiveOS LiMid will be evaluated by comparing with the EJVM mechanism (simpleRTJ), the DLM mechanism (Contiki DLM), and the Atmel OTAU mechanism (a reprogramming mechanism which does not use the middleware to decouple the application code from the system code) [24] from the perspectives of the code memory size, the application execution efficiency, and the reprogramming performance.

The code memory size of different mechanisms is shown in Figure 3.12. In LiveOS, the pre-linking mechanism is used, and most of the operations are performed on the personal computer rather than on the WSN nodes. As a result, the code memory cost is lower if compared to the other mechanisms.

The application code execution efficiency will have a direct influence on the energy cost of the WSN nodes. In the DLM and the LiMid mechanisms, the executable application code is the machine code. In the EJVM mechanism, the executable application code is the bytecode. To perform the comparison between these two kinds of code, the time between the invoking of the application system-call function and the execution of the corresponded system service function in the EJVM simpleRTJ and the Contiki DLM is calculated respectively, and the results show that the execution time of the simpleRTJ is 34.6 times that of the Contiki. This result proves that the EJVM which uses the bytecode has the execution efficiency much lower than the DLM which uses the machine code.

Application reprogramming performance is another critical evaluation standard for the middleware, and this can be evaluated by the reprogramming code size. In Figure 3.12, the reprogramming code sizes of the different mechanisms are shown. The Atmel OTAU does not implement the middleware concept. As a result, the monolithic software needs to be reprogrammed and the reprogramming code size is large. The LiMid uses the pre-linked machine code and the application code size is the smallest. As shown in the results, the reprogramming code size of the LiMid has respectively 92.7 per cent, 76.6 per cent, and 99.8 per cent been optimized if compared to that of the simpleRTJ, the Contiki DLM, and the Atmel OTAU mechanisms.

From the evaluations above, it is shown that the memory cost and the energy cost of LiMid are low, whereas the reprogramming performance is high. Due to

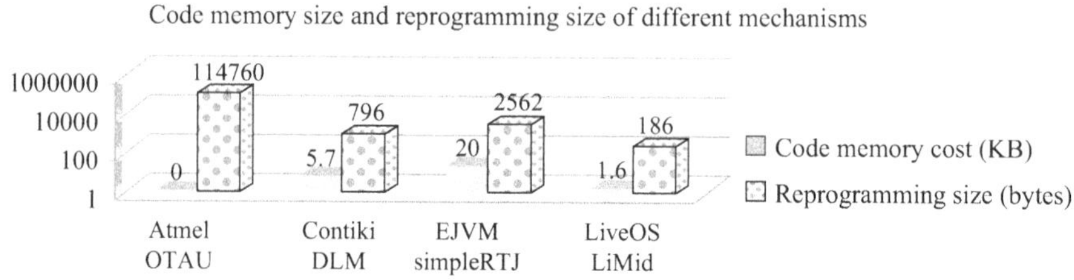

Figure 3.12 Code memory size and application reprogramming size of the different middleware mechanisms. A basic packet transmission program is used for this evaluation

this reason, LiMid is appropriate to run even on the tightly resource-constrained WSN nodes and can provide a friendly application programming and reprogramming environment to the WSN users.

3.5 LiveOS multi-core task assignment for the energy conservation

Energy conservation is significant for the WSN not only because most WSN nodes are equipped with the limited energy resources, but also because the recharging to the nodes after the deployment is commonly difficult. Currently, a set of energy conservation mechanisms have been implemented, such as the data prediction mechanism which can reduce the sampling redundancy of the sensor devices [25,26], the data compression and data aggregation mechanisms which can optimize the size of the packet to be transmitted [27–30], the topology control technique which can reduce the communication redundancy of the network [31–34], and the network protocols which can shorten the active period of the wireless transceivers [35–39]. These mechanisms are effective and can prolong the lifetime of the nodes. Yet, the energy limitation is still a challenge in the WSN and needs to be addressed further.

In this section, a new LiveOS energy conservation approach will be presented. Different from the other conservation mechanisms which focus mainly on the software aspect, the LiveOS energy conservation is realized by combining both the software technique and the multi-core hardware technique. With the support of the multi-core hardware platform, the energy cost of the WSN nodes can be optimized further.

3.5.1 Concept of the LiveOS multi-core energy conservation mechanism

The LiveOS multi-core energy conservation approach is based on the experimental results that the energy cost of executing one task on the different microcontrollers can be different (Table 3.2). Thus, several feature-different microcontrollers can be equipped on the WSN nodes. During the run-time, a WSN task can be assigned to the

Table 3.2 Energy and time cost of executing the tasks on different microcontrollers

Tasks for the measurements (voltage: 3V)	32-bit ARM AT91SAM7Sx			8-bit AVR ATmega1281		
	Current (mA)	Time (ms)	Energy (mJ)	Current (mA)	Time (ms)	Energy (mJ)
Temperature light, and humidity sensor sampling task	22.1	1386	91.89	15.9	1560	74.4
Signal processing task (pure instruction execution)	19.7	5.0	0.296	9.9	268	7.96
Flash programming (100-byte)	20.9	58	3636	16.3	103	5.037
Wireless packet transmission	21.8	139	9.091	20.9	132	8.276
Sleep	0.2	N/A	N/A	40 μA	N/A	N/A

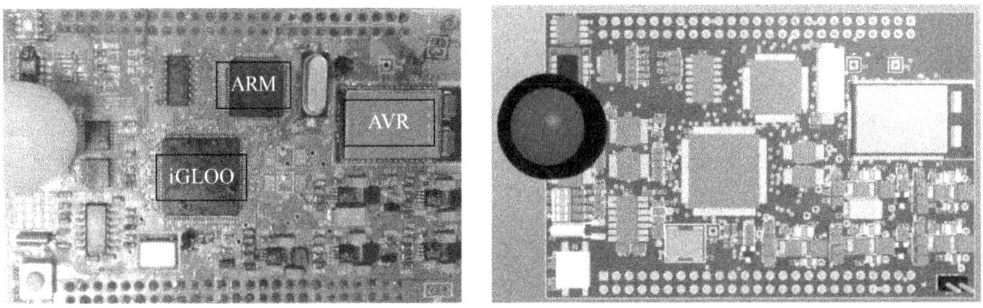

Figure 3.13 Multi-core EMWSN node

microcontroller which is the most energy-efficient to execute this task. By so doing, the energy cost of the WSN nodes can be optimized efficiently.

Currently, the LiveOS multi-core energy conservation approach has been implemented on the multi-core EMWSN node (Figure 3.13). EMWSN is composed of three microcontrollers: the 8-bit low-power AVR Atmega1281 [40], the 32-bit ARM AT91SAM7x [41], and the low-power iGLOO nano FPGA [42] (Figure 3.14). The iGLOO FPGA is configurable. With its configurability functionality, the working modes of the AVR and ARM microcontrollers (active, sleeping, or power off) can be adjusted without the necessity of changing the wired connection. By the above means, the WSN nodes can be energy aware to the tasks, and the energy cost of the nodes can be optimized.

3.5.2 Performance evaluation

To evaluate the performance of LiveOS multi-core energy conservation mechanism, the energy cost of the multi-core EMWSN node, the single-core iLive node which is equipped with only the AVR Atmega1281 microcontroller, and the single-core

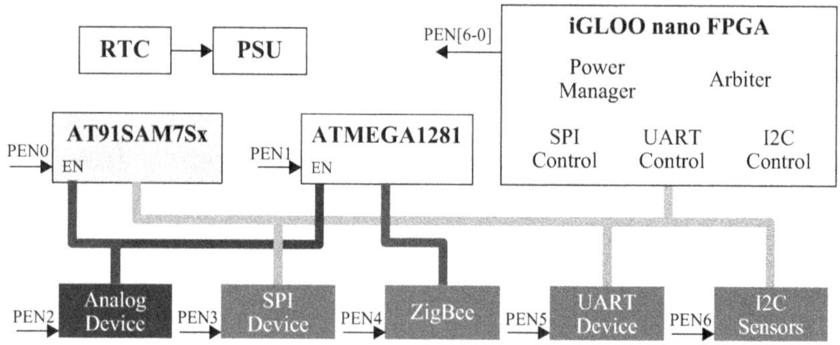

Figure 3.14 Circuit diagram of the multi-core EMWSN node

Live node which is equipped with only the ARM AT91SAM7x microcontroller are evaluated. In case the nodes are powered by a pair of AA lithium/iron disulfide (Li/FeS_2) 3,000 mAh batteries, and perform the temperature and light sampling task every 3 min, the lifetime of the EMWSN node, the iLive node, and the Live node will, respectively, be 1276 days, 829 days, and 382 days [43]. This result indicates that the energy cost of LiveOS multi-core EMWSN node can be respectively 35.1 per cent and 70 per cent optimized if compared to the traditional single-core iLive and Live node.

3.6 LiveOS multi-core task assignment to improve the real-time performance

In LiveOS, the multi-core technique has also been used to address the challenge of high-overhead real-time scheduling.

To achieve the real-time scheduling, a real-time algorithm needs to be implemented. Yet, each real-time algorithm has a schedulability condition, e.g., the schedulability condition of the rate-monotonic scheduling (RMS) algorithm is as follows [16]:

$$U = \sum_{i=1}^{k} \frac{C_i}{T_i} \leq k(2^{1/k} - 1) \tag{3.1}$$

where U represents the CPU utilization of the real-time tasks, C_i represents the task computation time, T_i represents the task release period, and k represents the number of real-time tasks to be scheduled. In case that two real-time tasks exist in the system ($k = 2$), the CPU utilization of these tasks will be limited to 82.8 per cent. Compared to the RMS, the Earliest Deadline First (EDF) algorithm supports higher CPU utilization rate, and it can be as high as 100 per cent [16,44]. Yet, no matter which algorithm is applied, the CPU utilization rate of the real-time tasks cannot exceed 100 per cent.

However, there can exist some cases in which the CPU utilization rate of the real-time tasks can be higher than 100 per cent, e.g., in Table 3.3, the utilization rate

Table 3.3 CPU utilization rate of different tasks on different microcontrollers

Tasks	Execution Period	On the AVR microcontroller		On the ARM microcontroller	
		Computation time	CPU utilization rate (per cent)	Computation time	CPU utilization rate (per cent)
Task1	20	5	25	3	15
Task2	10	6	60	5	50
Task3	25	8	32	6	24

of all the three tasks on the AVR microcontroller is 117 per cent. In this case, these tasks cannot be schedulable no matter what kind of real-time scheduling algorithm is used. However, if these tasks are distributed and executed concurrently on more than one microcontroller, they can be schedulable. In case that both the AVR and the ARM microcontrollers are equipped on the LiveOS multi-core node, and Task1 and Task3 are executed on the AVR microcontroller while Task2 is executed on the ARM microcontroller. Then, the CPU utilization rate of the tasks on the two microcontrollers will respectively be 57 per cent and 50 per cent. In this case, these tasks can be schedulable either by the RMS algorithm and the EDF algorithm.

3.7 LiveOS multi-core technique for the context-aware applications

With the development of WSN technology, WMSN application starts to emerge [45] and be merged with the traditional WSN application. One typical example is the forest fire detection application. In this application, two kinds of tasks exist: one is the temperature sampling task and the other is the fire image capturing task. Most of the time, only the temperature sampling task needs to be executed. When the sampled temperature is observed to be abnormal, e.g., the temperature is higher than 100°, the fire hazard may occur. Only in this case, the image capturing task needs to be launched so as to oversight the fire scene situation. In this example, the temperature sampling operation is a traditional WSN application. This application is simple and can be performed by the low-end microcontrollers. Yet, the image capturing operation is the WMSN application. This application is more complicated and needs to be performed by the powerful microcontrollers. Namely, the two feature-different tasks which have different requirements to the WSN hardware platforms are merged in this application.

If the single-core WSN node is used to execute the above applications, the powerful microcontroller needs to be equipped. In this case, this powerful microcontroller will not only perform the complicated image capture task, but also perform the simple temperature sampling task. Yet, it is high energy cost to use the powerful

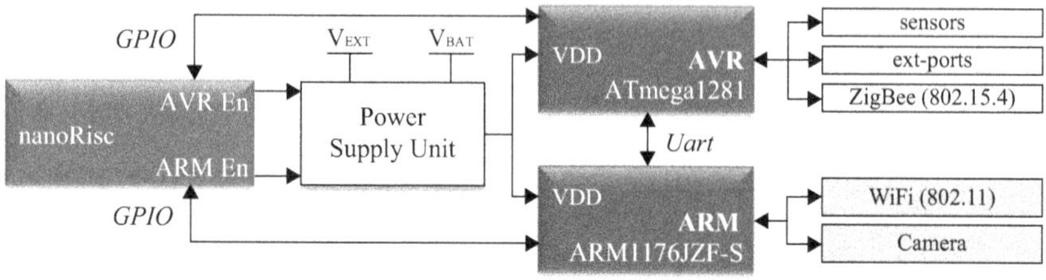

Figure 3.15 Elementary diagram of multi-core MiLive node. The working modes of AVR and ARM microcontrollers can be configured by the nanoRisc microcontroller through the power supply unit

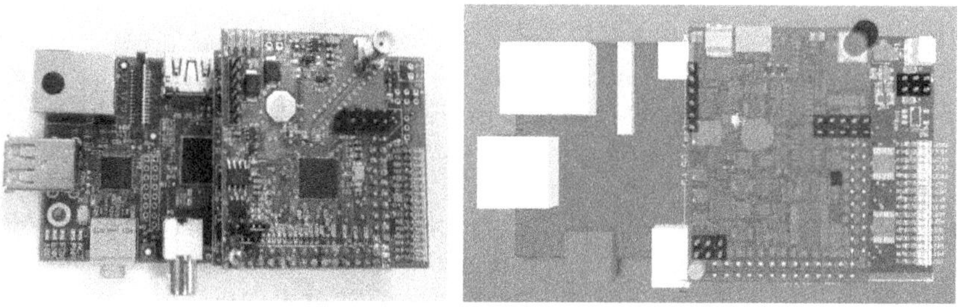

Figure 3.16 Multi-core MiLive node. Both the WSN microcontroller and the WMSN microcontroller are equipped so that the node can be aware to different application contexts

microcontroller to run the simple temperature sampling task. As a result, more energy cost will be consumed, and the lifetime of the node will be shortened.

To address the challenge above, the LiveOS multi-core application context-aware mechanism is implemented. With this mechanism, three kinds of microcontrollers will be equipped on the LiveOS multi-core node MiLive (Figures 3.15 and 3.16). One is the low-power 8-bit AVR microcontroller ATmega1281 which will be used dedicatedly to perform the temperature sampling task, one is the high-power 32-bit ARM microcontroller ARM1176JZF-S which will be used dedicatedly to perform the image capturing task, and another is the ultra-low-power 4-bit auxiliary microcontroller nanoRisc which will be used to configure the working modes of the AVR and ARM microcontrollers. Most of the time, only the AVR microcontroller works and the environmental temperature is monitored. Only when the sampled temperature value exceeds the warning level, the ARM microcontroller is powered on to startup the scene image capturing work. By doing so, the WSN node can become context aware to the different applications and the node platform resources can be utilized more reasonably. Currently, an online live demo about the MiLive node is available from the website: http://edss.isima.fr/demoforall/. (login name: demo, password: demo).

3.8 LiveOS multi-core fault-tolerant mechanism

Fault tolerance is significant for the WSN as the WSN nodes can be deployed in the harsh environments where they are difficult to be maintained after being deployed. With the fault-tolerant functionality, the economic loss and the maintenance cost caused by the failed nodes can be decreased.

Currently, a set of fault-tolerant mechanisms have been realized in the WSN, such as the multipath routing mechanism which can provide active route replication for the network [46], the topology management mechanism which can recover the communication fail from the network reconstruction [47], and the data aggregation mechanism which can still provide sufficient information to the recipients even if parts of the nodes are failed [29]. All the mechanisms are effective to improve the availability of the network. Yet, they are limited in that they focus mainly on recovering the network from the faults and lacking in the research of recovering the nodes from the fail. To improve the fault-tolerant ability of the WSN nodes, a new LiveOS multi-core fault-tolerant technique is implemented and will be presented in this section.

3.8.1 Concept and implementation of the LiveOS multi-core fault-tolerant platform

Traditionally, the WSN nodes are the single-core nodes and only one microcontroller is equipped. In case this microcontroller runs abnormally, the node will fail. To improve the fault-tolerant performance, two microcontrollers can be equipped on the WSN nodes: the working microcontroller and the auxiliary monitoring microcontroller. The working microcontroller takes charge of performing the WSN tasks whereas the monitoring microcontroller takes the responsibility of monitoring the run-time status of the working microcontroller. In case the working microcontroller is observed to run abnormally, the monitoring microcontroller will take action to recover it. By this means, the WSN node can recover from the faults and continue its functioning.

Currently, the above LiveOS multi-core fault-tolerant mechanism has been achieved on the multi-core IWoT node. IWoT node is equipped with the 8-bit working microcontroller AVR ATmega1281 and the 4-bit ultra-low-power high-reliable auxiliary microcontroller nanoRisc. During the run-time, the working microcontroller will send the execution status information to the auxiliary core at the time. In terms of the received information, the auxiliary microcontroller can diagnose whether the working microcontroller runs normally or not. If the received status information is incorrect or it is not received within the expected time, the working microcontroller can be determined to be abnormal. In this case, the auxiliary microcontroller will restart the working microcontroller through the power supply unit.

A key design principle of the above LiveOS multi-core fault-tolerant mechanism is that the auxiliary microcontroller should be a high reliable. Otherwise, the fault recovery cannot be performed ideally. Fortunately, this design objective is not difficult to be achieved. On the one hand, the software running on the auxiliary microcontroller

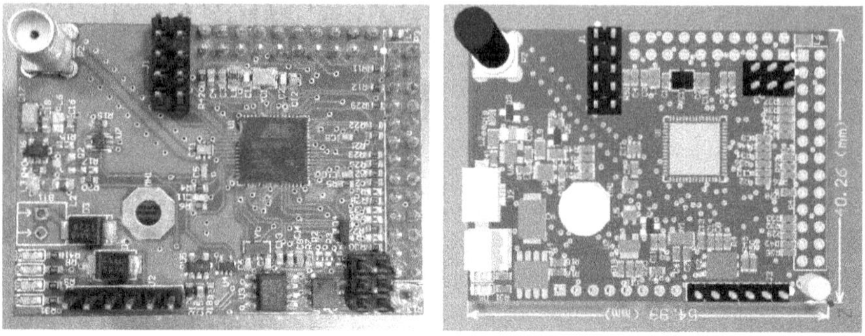

Figure 3.17 LiveOS fault-tolerant multi-core IWoT node

is quite simple, thus the software reliability can be high. On the other hand, a high-reliable auxiliary microcontroller can be selected during the hardware design process. By doing this, the hardware platform of the auxiliary microcontroller can be high. In case that the above two conditions are satisfied well, the reliability of the auxiliary microcontroller can keep well.

3.8.2 Experimental evaluation

To evaluate the fault-tolerant ability of the LiveOS multi-core node, three multi-core IWoT nodes (Figure 3.17) were deployed in the outdoor garden of the ISIMA more than 3 years ago. By experiment, these nodes ran abnormally during the past period, yet they can be recovered automatically. As a result, no nodes have been failed until now. However, before the LiveOS multi-core technique is applied, more than 30 per cent of the traditional single-core Live node [44] will fail after being deployed for about 2 months. Currently, an online demo which can be used to verify the reliability of the multi-core IWoT node is available from the website: edss.isima.fr (login name: demo, password: demo).

3.9 LiveOS multi-core debugging mechanism

Compared to the debugging on the general-purpose platforms, the debugging on the WSN platform is more difficult. First, the WSN platform resources are constrained, thus the debugging approach in the WSN needs to be low overhead. Second, the WSN nodes commonly work within a network, rather than work in isolation. Thus, the run-time process of the WSN nodes cannot be halted during the debugging process. Once being halted, the node will stop exchanging with the others. And then, it can be regarded as a lost one and be deleted from the network. Due to the above reasons, the development of a new debugging approach which can be used to debug the resource-constrained WSN network effectively becomes essential.

3.9.1 Traditional debugging approaches

The traditional debugging approaches which can be applied on the embedded platforms include the *printf*, the serial port and the breakpoint. The former two approaches are not effective in the WSN as the debugging overhead is high. In these cases, a passive influence will be caused to the regular execution of the WSN code. Breakpoint approach is also not appropriate for the WSN debugging. First, it is more effective if the software execution is logically sequential, but not effective when the software execution is concurrent. Second, it is more effective to debug an isolated WSN node, but not effective to debug a WSN network as the nodes can be deleted from the network after the breakpoint is met.

3.9.2 Concept and implementation of the LiveOS multi-core debugging approach

To address the challenge of debugging on the resource-constrained WSN platform, the LiveOS multi-core debugging mechanism is implemented. The concept of the LiveOS multi-core debugging approach is to use an assistant resource-abundant debugging microcontroller to ease the debugging overhead on the resource-constrained WSN microcontroller. With the assist of the debugging microcontroller, most debugging tasks which should originally perform by the WSN microcontroller will be migrated to the debugging microcontroller. By so doing, the debugging overhead on the resource-constrained WSN microcontroller can become light and the WSN debugging challenges described above can be addressed soundly.

Currently, the LiveOS multi-core debugging mechanism has been realized on the MiWSN node. MiWSN node is equipped with two microcontrollers: the resource-constrained WSN microcontroller AVR ATmega1281 and the resource-abundant debugging microcontroller ARM1176JZF-S. Two microcontrollers communicate with each other through the general-purpose input and output (GPIO) ports (Figure 3.18). GPIO is selected as the communication port since it has low communication overhead. With this multi-core architecture, the debugging tasks on the WSN microcontroller only involve the production of the debugging code and the transmit of the debugging code from the GPIO ports to the debugging microcontroller. As for the left debugging tasks such as the caching of the produced

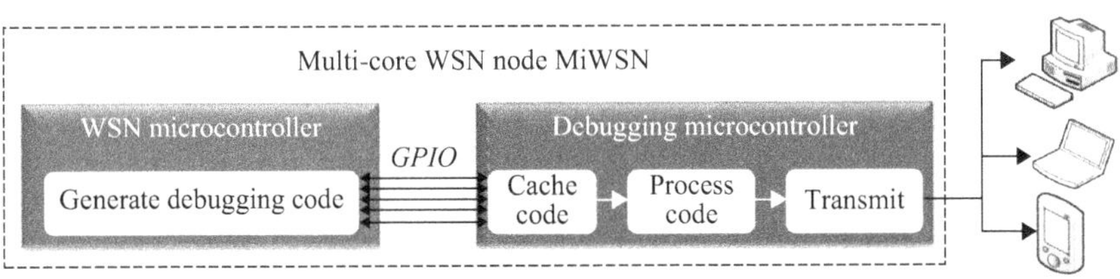

Figure 3.18 LiveOS multi-core debugging system. With the assist of powerful debugging microcontroller, the debugging overhead on the WSN microcontroller can be optimized

debugging code, the analysis and processing of the debugging information, they will be performed by the powerful debugging microcontroller. Currently, this multi-core debugging approach has been applied to debug the scheduling process of the hybrid LiveOS, and an available online demo about this can be accessed from the website: http://edss.isima.fr/sites/smir/miros.

3.10 Discussion on the LiveOS design concepts

LiveOS is different from the other OSes in that it addresses a set of critical WSN challenges by combining both the software technique and the multi-core hardware technique, rather than only from the software aspect. The evaluation results on the LiveOS prove that this research way is feasible and effective. Although several micro-controllers are equipped on the multi-core nodes and the node price will increase in a way, it is not a considerable matter since the WSN microcontrollers are commonly low-cost one, e.g., the cost of the AVR Atmega1281 can be lower than $5.

Each approach has advantages and drawbacks. It cannot exist an approach which has the best performance in all the evaluation aspects. Yet, it can exist an approach which is the most appropriate for a given situation, e.g., the shared-stack multithreading mechanism shows the advantage in low memory consumption, yet it is achieved at the cost of high scheduling overhead. Thus, it is appropriate to be used when the memory resource of the WSN platform is high constrained, whereas the energy resource of the WSN node is abundant.

The design concept of some mechanisms is not complicated, yet this does not indicate that these approaches are not essential, e.g., the software architecture of the pre-linked machine-code LiMid is not so complicated as that of the EJVM and the dynamic relocation schemes, but it is typically appropriate to be used on the high resource-constrained WSN nodes.

Different multi-core mechanisms can be integrated to improve the WSN performance comprehensively, e.g., the LiveOS multi-core energy conservation scheme and the LiveOS multi-core fault-tolerant approach can be applied comprehensively on the one WSN node. By doing this, the performance of the WSN platform can be improved comprehensively.

3.11 Conclusions and ongoing works

LiveOS is a new real-time, memory-efficient, energy-efficient, user-friendly, context-aware and fault-tolerant WSN OS. It implements several new design mechanisms, such as the hybrid scheduling, the shared-stack multithreading, the reactive-defragmentation allocation, the resource-efficient middleware LiMid, the multi-core energy conservation, the multi-core real-time responsiveness, the multi-core context-aware ability, the multi-core fault-tolerant functionality, and the multi-core WSN debugging. By means of these mechanisms, the stack memory cost of the LiveOS can be more than 40 per cent decreased if compared to the traditional multithreaded

MantisOS; the energy cost of LiveOS can be more than 30 per cent reduced if compared to the traditional single-core WSN system; the real-time performance of LiveOS can be more than 16 per cent optimized if compared to the single-core WSN node. In addition, the user application programming can be simplified, the application reprogramming performance can be improved, the WSN/WMSN integrated application context-aware ability can be improved and the WSN fault-tolerant ability can be enhanced. Due to the above features, LiveOS becomes feasible to run on the resource-constrained WSN nodes, can support the real-time outdoor WSN applications and can provide a friendly development environment to the users. For more works about the team works, it can be accessed from the homepage: http://edss.isima.fr/sites/smir/.

The ongoing works of LiveOS will involve the following aspects. First, the research of a multi-core task scheduling mechanism which can optimize both the energy cost and the real-time performance of the WSN nodes will be done. Currently, the multi-core technique has been used either to optimize the energy cost or to optimize the real-time performance. In the next step, the multi-core scheduling mechanism which can optimize the energy cost and the real-time performance simultaneously will be investigated. Since the energy optimization condition and the real-time optimization condition can conflict with each other commonly, a multi-objective optimization scheduling algorithm needs to be researched. Second, the design and implementation of a comprehensive multi-core WSN node will be realized. Currently, different kinds of multi-core nodes have been developed so as to satisfy the requirements of different contexts. This development way has complicated manufacture process and high maintenance cost. In the ongoing works, a comprehensive multi-core platform which can self-configure to adapt to the different application contexts will be designed and implemented. With this platform, the WSN nodes can be more intelligent and the WSN development cycle can also be shorten.

Acknowledgments

The authors would like to thank all the colleagues and copartners who have contributed to this study, and are also grateful for the research support from the Natural Science Foundation of HUAT (BK201411), the Sci-Tech. Pillar program of Hubei Province (2014BHE024), the National High-tech 863 R&D Program of China (2015AA015403), the fundamental research funds for the Central Universities (WUT:163110003, WUT:40120225), and the National Key Technology R&D Program (2012BAH45B01).

References

[1] El Emary I.M.M., Ramakrishnan S. *Wireless Sensor Networks: From Theory to Applications*. CRC Press; 2013.

[2] Akyildiz I.F., Vuran M.C. *Wireless Sensor Networks*. John Wiley & Sons Press; 2010.

[3] Borges L.M., Velez F.J., Lebres A.S. 'Survey on the characterization and classification of wireless sensor network applications'. *IEEE Communications Surveys and Tutorials*. 2014;16(4):1860–1890.

[4] Texas Instruments Inc. *The internet of things: Opportunities & Challenges* [online]. Available from http://www.ti.com/ww/en/internet_of_things/iot-challenges.html [Accessed 26 March 2016]

[5] Wikipedia. *List of Wireless Sensor Nodes* [online]. Available from https://en.wikipedia.org/wiki/List_of_wireless_sensor_nodes [Accessed 26 March 2016]

[6] Hill J., Szewczyk R., Woo A., Hollar S., Culler D., Pister K. 'System architecture directions for networked sensors'. In *Proceedings of the 9th International Conference on Architectural Support for Programming Languages and Operating Systems*; Cambridge, USA, 12–15 November 2000. pp. 93–104.

[7] Dunkels A., Gronvall B., Voigt T. 'Contiki—a lightweight and flexible operating system for tiny networked sensors'. In *Proceedings of Local Computer Networks*; Tampa, FL, USA, 16–18 November 2004. pp. 455–462.

[8] Han C., Kumar R., Shea R., Kohler E., Srivastava M. 'A dynamic operating system for sensor nodes'. In *Proceedings of Mobile Systems, Applications, and Services*; Seattle, WA, USA, 6–8 June 2005. pp. 117–124.

[9] Bhatti S., Carlson J., Dai H., Deng J. 'MANTIS OS: an embedded multithreaded operating system for wireless micro sensor platforms'. *ACM Kluwer Mobile Networks and Applications*. 2005;10(4):563–579.

[10] Watteyne T., Vilajosana X., Kerkez B., Chraim F. 'OpenWSN: a standards-based low-power wireless development environment'. *Transactions on Emerging Telecommunications Technologies*. 2012;23(5):480–493.

[11] Chen Y.T., Chien T.C., Chou P.H. 'Enix: a lightweight dynamic operating system for tightly constrained wireless sensor platforms'. In *Proceedings of the 8th ACM Conference on Embedded Networked Sensor Systems*; Zurich, Switzerland, 3–5 November, 2010. pp. 183–196.

[12] Cao Q., Abdelzaher T., Stankovic J., He T. 'The LiteOS operating system: towards UNIX-like abstractions for wireless sensor networks'. *In International Conference on Information Processing in Sensor Networks*; St. Louis, Missouri, USA, 22–24 April, 2008. pp. 233–244.

[13] Ousterhout J. 'Why threads are a bad idea (for most purposes)'. In *Proceedings of the 1996 USENIX Annual Technical Conference*; San Diego, CA, USA, 22–26 January 1996. Volume 5.

[14] Von Behren R., Condit J., Brewer E. 'Why events are a bad idea (for high-concurrency servers)'. In *Proceedings of the 9th Workshop on Hot Topics in Operating Systems*; Lihue, HI, USA, 18–21 May 2003. pp. 19–24.

[15] Stephen Ferg. *Event-Driven Programming: Introduction, Tutorial, History* [online]. 2006. Available from http://eventdrivenpgm.sourceforge.net [Accessed 26 March 2016]

[16] Liu C.L., Layland J.W. 'Scheduling algorithms for multiprogramming in a hard-real-time environment'. *Journal of the ACM*. 1973;20(1):46–61.

[17] Klues K., Liang C.J.M., Paek J., *et al.* 'TOSThreads: thread-safe and non-invasive preemption in TinyOS'. In *Proceedings of the ACM Conference on Embedded Networked Sensor Systems*; Berkeley, CA, USA, 4–6 November 2009. pp. 127–140.

[18] Wilson P.R., Johnstone M.S., Neely M., Boles D. 'Dynamic storage allocation: a survey and critical review'. *Memory Management*. Springer Berlin Heidelberg, 1995. pp. 1–116.

[19] Adam D. *Contiki Managed Memory Allocator* [online]. 2012. Available from https://github.com/adamdunkels/contiki-fork/wiki/Memory-allocation [Accessed 26 March 2016]

[20] RTJ Computing Pty. Ltd. *simpleRTJ Technical Brief* [online]. 2009. Available from http://www.rtjcom.com/downfile.php?f=techpdf [Accessed 26 March 2016]

[21] LEGO Mindstorms. *LeJOS, Java for Lego Mindstroms* [online]. 2006. Available from http://www.lejos.org [Accessed 26 March 2016]

[22] Tills Palm Pages. *The NanoVM – Java for the AVR* [online]. 2005. Available from http://www.harbaum.org/till/nanovm/index.shtml [Accessed 26 March 2016]

[23] Brouwers N., Langendoen K., Corke P. 'Darjeeling, a Feature-Rich VM for the Resource Poor'. *In Proceedings of Embedded Networked Sensor System*; Berkeley, CA, USA, 4–6 November 2009. pp. 169–182

[24] Atmel Corporation. *Atmel AVR2058: BitCloud OTAU User Guide* [online]. 2015. Available from http://www.atmel.com/Images/Atmel-8426-BitCloud-OTAU_User-Guide_AVR2058.pdf [Accessed 26 March 2016].

[25] Kanagal B., Deshpande A. 'Online filtering, smoothing and probabilistic modeling of streaming data'. In *Proceedings of the 24th International Conference on Data Engineering*; Cancún, México, April 2008. pp. 1160–1169.

[26] Jain A., Chang E.Y., Wang Y.F. 'Adaptive stream resource management using Kalman filters'. In *Proceedings of the ACM International Conference on Management of Data*; Paris France, 13–18 June, 2004. pp. 11–22.

[27] Caione C., Brunelli D., Benini L. 'Distributed compressive sampling for lifetime optimization in dense wireless sensor networks'. *IEEE Transactions on Industrial Informatics*. 2012;8(1):30–40.

[28] Srisooksai T., Keamarungsi K., Lamsrichan P., Araki K. 'Practical data compression in wireless sensor networks: A survey'. *Journal of Network and Computer Applications*. 2012;35(1):37–59.

[29] Maraiya K., Kant K., Gupta N. 'Wireless sensor network: a review on data aggregation'. *International Journal of Scientific and Engineering Research*. 2011;2(4):1–6.

[30] Xiang L., Luo J., Vasilakos A. 'Compressed data aggregation for energy efficient wireless sensor networks'. In *Proceedings of IEEE Conference on Sensor, Mesh and Ad hoc Communications and Networks*; Salt Lake, Utah, USA, 27–30 June, 2011. pp. 46–54.

[31] Aziz A.A., Sekercioglu Y.A., Fitzpatrick P., Ivanovich M. 'A survey on distributed topology control techniques for extending the lifetime of battery

powered wireless sensor networks'. *Communications Surveys and Tutorials*. 2013;15(1):121–144.

[32] Li M., Li Z., Vasilakos A.V. 'A survey on topology control in wireless sensor networks: taxonomy, comparative study, and open issues'. *Proceedings of the IEEE*. 2013;101(12):2538–2557.

[33] Üster H., Lin H. 'Integrated topology control and routing in wireless sensor networks for prolonged network lifetime'. *Ad Hoc Networks*. 2011;9(5): 835–851.

[34] Warrier A., Park S., Mina J. and Rheea I. 'How much energy saving does topology control offer for wireless sensor networks? – a practical study'. *ACM Computer Communications*. 2007;30(14):2867–2879.

[35] IEEE Standards Association. *IEEE SA-802.15.4-2006-IEEE Standard for Information Technology* [online]. Available from https://standards.ieee.org/findstds/standard/802.15.4-2006.html [Accessed on 26 March 2016]

[36] Saleh A.M.S., Ali B.M., Rasid M.F.A., Ismail A. 'A survey on energy awareness mechanisms in routing protocols for wireless sensor networks using optimization methods'. *Transactions on Emerging Telecommunications Technologies*. 2014;25(12):1184–1207.

[37] Gilani M.H.S., Sarrafi I., Abbaspour M. 'An adaptive CSMA/TDMA hybrid MAC for energy and throughput improvement of wireless sensor networks'. *Ad Hoc Networks*. 2013;11(4):1297–1304.

[38] Zareei M., Taghizadeh A., Budiarto R., Wan T.C. 'EMS-MAC: energy efficient contention-based medium access control protocol for mobile sensor networks'. *Computer Journal*. 2011;54(12):1963–1972.

[39] Pantazis N., Nikolidakis S.A., Vergados D.D. 'Energy-efficient routing protocols in wireless sensor networks: a survey'. *IEEE Communications Surveys and Tutorials*. 2013;15(2):551–591.

[40] Atmel Corporation. *8-bit Microcontroller with 64K/128K/256K Bytes In-System Programmable Flash* [online]. 2007. Available from www.atmel.com/Images/doc0945.pdf [Accessed 26 March 2016]

[41] Atmel Corporation. *Atmel SAM7X Series of Microcontrollers* [online]. 2014. Available from http://www.atmel.com/Images/doc6120.pdf [Accessed 26 March 2016]

[42] Microsemi Corporation. *IGLOO nano FPGAs* [online]. 2016. Available from http://www.microsemi.com/products/fpga-soc/fpga/igloo-nano [Accessed 26 March 2016]

[43] Shi H.L. *Development of an energy efficient, robust and modular multi-core wireless sensor network*. PhD thesis of University Blaise Pascal, 2014.

[44] Buttazzo G.C. 'Rate monotonic vs. EDF: judgment day'. *Real-Time Systems*. 2005;29(1):5–26.

[45] Akyildiz I.F., Melodia T., Chowdhury K.R. 'A survey on wireless multimedia sensor networks'. *Computer Networks*. 2007;51(4):921–960.

[46] Chanak P., Banerjee I. 'Energy efficient fault-tolerant multipath routing scheme for wireless sensor networks'. *Journal of China Universities of Posts and Telecommunications*. 2013;20(6):42–61.

[47] Younis M., Senturk I.F., Akkaya K., Lee S., Senel F. 'Topology management techniques for tolerating node failures in wireless sensor networks: a survey'. *Computer Networks*. 2014;58:254–283.

Chapter 4
OSIRIS framework: sensOr-baSed monItoRIng Systems*

Raphael Guerra[1] *and Felipe Santos*[1]

Abstract

Deploying monitoring systems with wireless sensor networks (WSNs) is a very challenging task: physical components are highly heterogeneous, suffer damage, replaced; data is generated massively and must be managed, stored, and made available to other systems. In this chapter, we propose OSIRIS, a framework for building monitoring systems based on WSNs. This framework provides resources for monitoring the WSN, collecting, processing, and storing data, and an interface for providing data to other applications and/or systems. OSIRIS uses a set of abstractions to offer flexibility for the creation of various monitoring systems and to decouple network physical sensors from data consuming applications. In our tests, we use an implementation of OSIRIS to show that our architectural decisions allow OSIRIS to handle commercial and industrial-sized networks.

4.1 Introduction

Wireless sensor networks (WSNs) consist of small sensors with limited computational resources working intelligently in groups to achieve their goals. These networks have the main purpose of offering to other applications environmental sensory data. The deployment of these networks in practical applications has, as biggest challenges energy restriction and low processing capacity [1]. Several applications of these networks have been proposed in the literature [2–4], including smart environments and Internet of Things [5–8].

Sharing collected data with other applications is an additional challenge to the implementation of systems based on WSN. Among the challenges for this interoperability, we can mention the highly heterogeneous nature of WSNs and the volatile

*Development repository at https://github.com/labtempo/osiris and binaries at https://github.com/labtempo/osiris-binaries

[1]Computation Institute, Federal Fluminense University, Niterói/RJ, Brazil, Email: {rguerra,fralph}@ic.uff.br

connectivity of nodes. Furthermore, network data traffic is the main source of energy consumption. These data often still need to be processed before being used to provide a more complete and robust view of the monitored environment, a technique known as data fusion [9]. In this context, an infrastructure becomes necessary to collect, process, store, and share data between multiple concurrent applications from different sources in a standardized and flexible way and manage the sensor network while hiding intrinsic details of how data are obtained. Building this infrastructure from scratch for each monitoring system is a costly task.

SenseWeb [10] is a framework that tries to address these problems. It defines four main components: *sense gateway*, *mobile proxy*, *coordinator*, and *data transformer*. *Sense gateways* provide a uniform interface for communication with the sensors. *Mobile proxies* act similar to *sense gateways* for mobile sensors: they return data from any device that meets the needs expressed by a requesting application. *Coordinator* stores data in cache to minimize the flow of data directly from the sensors and manages each sensing application needs to locate appropriate sensors. Finally, *Data transformers* handle the data collected before passing them on to applications, e.g., for better viewing, unit conversion, or filtering.

Data transformers require applications to be aware of the need to perform data transformations. Depending on how data transformers are implemented, raw data that must be handled before passed on to the applications may be processed at each request, increasing processing demand and latency. Moreover, if two applications request different data from the same sensor network, there might be data traffic in the network for each request, increasing energy consumption. We also noticed that the *coordinator* is a centralizing element, compromising scalability. Finally, SenseWeb does not offer an element to manage the operation of WSNs, such as energy consumption of nodes and state of routing trees.

Sensor Cloud [8] is another proposal of a framework to support the construction of sensor-based monitoring systems. Sensor Cloud focuses on abstracting data from their sources using virtual nodes [11] and managing access to these data in a cloud infrastructure.

Both SenseWeb and Sensor Cloud focus primarily on building an information management infrastructure to collect, group, and share data from several WSNs. The open question is how to extract data from the sensor network to the information management infrastructure while addressing issues such as reduce network traffic, handle sensor heterogeneity and provide fault tolerance.

We propose the OSIRIS Framework: sensOr-baSed monItoRIng Systems. As can be seen in Figure 4.1, it consists of five main modules: *Collector*, *SensorNet*, *VirtualSensorNet*, *Function*, and *External*. *Collector* is the gateway that extracts data from the sensor network and transfers them to OSIRIS; there may be several of them in the same network so that data, have multiple options to leave the sensor network. *SensorNet* is the software representation of the current state of the physical network and its sensors, storing data and metadata such as battery level, current reading of each sensor, and network routing graph. *VirtualSensorNet* is the entity that manages the virtual sensors, an abstraction that hides data sources from applications. Data source can be a physical sensor or a software element that processes data from one or more sources (for fault tolerance, data fusion, etc.). Each module *Function* is

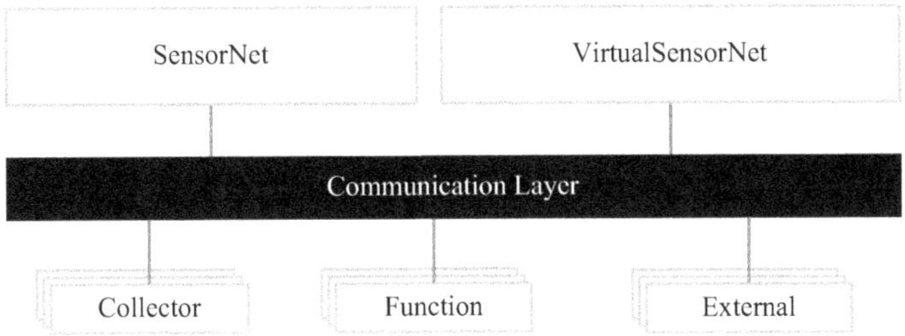

Figure 4.1 OSIRIS architecture

a shared module that works linked to virtual sensors in order to process data and return a result. Application-specific modules to display data or forward data to other subsystems are implemented as *External* modules. All these modules are distributed and communicate with each other over a *Communication Layer*.

SensorNet and *VirtualSensorNet* prevent successive data queries from generating extra traffic on the physical sensor network, saving energy. In addition, *VirtualSensorNet* and *Function* together allow software processed data to be stored and shared as if they came directly from sensors. This abstraction reduces processing load on the framework and coupling between applications and physical data sources. The strategic placement of *Collectors* provides for fault tolerance and reduced traffic and energy consumption within the network by offering multiple exit routing options for collected data [12]. Since this module is distributed, its instances can runon embedded devices and be distributed over a wide area. *Collectors* also directly handle redundancy.

Our experiments show that our implementation of OSIRIS supports sensor networks with traffic rate up to 10.000 samples per second. Given that even commercial applications of WSN tend to use sample frequency in the order of few minutes to reduce data traffic, and hence save energy [4,13,14], this throughput suffices for WSNs with over 1 million nodes. This validates our architectural decisions with respect to scalability (*Collectors* are distributed) and efficiency. OSIRIS's capability to support the implementation of sensor-based monitoring systems has been demonstrated in previous work [15].

The rest of this chapter is structured as follows. Section 4.2 describes the communication layer and Section 4.3, OSIRIS modules. Then, we present in Section 4.4 experimental results about the performance and efficacy of OSIRIS. Finally, Section 4.5 brings our concluding remarks.

4.2 OSIRIS Communication Layer

We developed a protocol for OSIRIS Communication Layer called OMCP – *OSIRIS module communication protocol*. In the next three subsections, we describe OMCP, the communication between modules, and OMCP's implementation on RabbitMQ [16].

4.2.1 OMCP Protocol

OMCP is a communication protocol designed to allow synchronous and asynchronous communication between distributed modules pairwise and via publish/subscribe. Inspired on HTTP, OMCP is a client/server protocol that implements REST architectural style [17]. This architectural decision aims at reduced network traffic when broadcasting sensor read samples, yet allowing pairwise communication with loose coupling between modules. This way, modules can be dynamically added to the framework when building monitoring system applications.

OMCP defines five request methods: GET, POST, PUT, DELETE e NOTIFY. The first four methods are used for synchronous communication: a client issues a request and waits for the server to process and send a response. They behave as in HTTP1.1 [18]: GET retrieves the state of a resource, POST creates a resource, PUT updates the state of a resource, and DELETE deletes a resource. NOTITY is an asynchronous method used to send data from one client to one or more servers.

A request packet, issued by clients, contains fields to indicate the method, the resource, the protocol version, and the destination address (module that will handle the request). A response packet contains the protocol version, a return code to indicate whether the request was properly processed or not, and a textual description of the return code. It also contains field to indicate the URI of a resource created at the server (for POST). Both packet types also contain fields for date and time at which the packet was issued, identification of the packet issuer, content type, content length, and payload.

4.2.2 OSIRIS modules communication

OSIRIS modules can communicate pairwise (point-to-point) synchronously and asynchronously, and broadcast event messages. Point-to-point communication is used so modules can consult, delete, or modify the state of resources in other modules. A module may also request the creation of resources in other modules. Broadcast communication is used to notify events to an undisclosed number of modules using publish/subscribe.

Broadcast communication is implemented using the concept of message groups. Producing modules publish events in groups of interest and consuming modules register to groups in other to get their messages. OSIRIS defines some standard message groups, see Section 4.3, and the developer is free to extend the framework with whatever groups necessary for correct application behavior.

OSIRIS also defines two communication components for modules: OMCP Client and OMCP Server. An OMCP Client may send request messages to an OMCP Server or publish to message groups. An OMCP Server may register to message groups and receive direct messages from OMCP Clients.

4.2.3 Implementation on RabbitMQ

We implemented OMCP on RabbitMQ [16], an open-source message broker middleware for distributed systems capable of running on server clusters and handling over 1 million messages per second [19]. It supports robust applications (atomic transactions, broadcast and multicast messages, delivery guarantee, etc.), is easy to use, available in several operating systems, and has a large community with considerable support for various development platforms. It implements the AMQP protocol [20], which is an open standard application layer protocol for message-oriented middleware.

The operation of RabbitMQ is simple and works with five basic elements [21]: *producer*, *consumer*, *queue*, *exchange*, and *binding*. *Producers* can send messages to an *exchange* or directly to a *queue*. *Exchanges* distribute copies of messages to zero or more queues using routing algorithms and rules called *bindings*. *Queues* pass message on to their respective *consumers* or *consumers* search for messages in them according to the demand.

Figure 4.2 depicts our implementation of OCMP on RabbitMQ. We implemented group messages with *exchanges*. Both OMCP Client and Server consists of a RabbitMQ producer, consumer, and queue. An OMCP Client sends out request packets

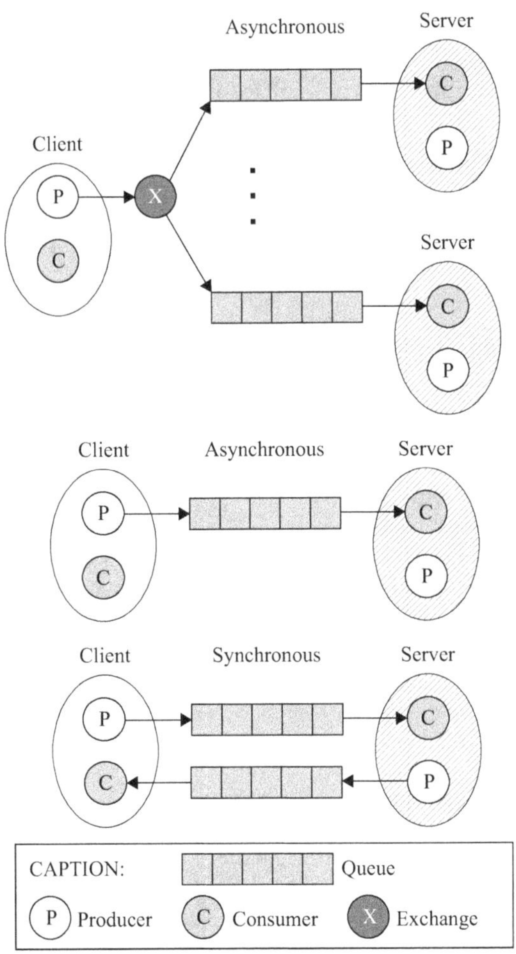

Figure 4.2 OMCP implementation on RabbitMQ

using its producer and, in case of synchronous messages, creates a temporary queue binded to its consumer in order to receive the response packet. Request packets of synchronous messages are mandatorily sent out to the queue of an OMCP Server, and request packets of asynchronous messages may be sent to either a queue binded to an OCMP server or to one or more exchanges. The OMCP Server has a consumer binded to its queue, and optionally to one or more exchanges, to receive request packets. In case of synchronous messages, it uses its producer to send out response packets to the temporary queue of the OMCP Client that issued the request. The OMCP Client informs the address of this temporary queue in a property field of the request packet.

In our current implementation, an OMCP Client waits indefinitely for a response from the OMCP Server. The server has a timeout for processing the request; after this timeout, the action is canceled and an error is reported to the client. In future updates, we want to implement Client timeout and Server rollback for reliable communication.

4.3 OSIRIS modules

Besides the communication layer, OSIRIS defines five module types: Collector, SensorNet, VirtualSensorNet, Function, and External. SensorNet and VirtualSensorNet have one unique instance each running at runtime and are offered as a service (daemon) to the framework user. Collector, Function, and External are modules that each developer has to implement specifically for his monitoring application. The next subsections describe each module in more detail.

4.3.1 Collector

Collector is responsible for collecting data from the sensor network, handling, and forwarding them to other modules of the framework. Arriving data have to be formatted and packetized in order to inform from which network, collector, and sensor they came. Each data is also formatted as a tuple with fields to specify *value, name, type, unit*, and *symbol*. For example, a temperature sensor read can be formatted as <*30.5, temperature, real, Celsius, C*>. All Collectors publish data to a standard message group dedicated for collector messages, called omcp://collector.messagegroup.osiris/.

Collector is a distributed module, *i.e.* there may be multiple instances running concurrently and distributed over the network, and have to be implemented each time a monitoring application is built using OSIRIS. Collectors are identified with IDs that may be unique or not. Collectors with same ID are seen as replica for fault tolerance in the sensor network, and data sampled at the same time and coming from the same network, collector, and sensor are stored only once in OSIRIS database.

4.3.2 SensorNet

SensorNet is responsible for monitoring the physical network and broadcasting notifications of network changes. It subscribes to message group omcp://collector.messagegroup.osiris/ and processes metadata informed by Collector modules to create an internal data structure with the following resources:

- *Networks* – Networks monitored by SensorNet.
- *Collectors* – Each monitored network consists of one or more Collectors. This resource is a list of Collectors for a given network.
- *Sensors* – Sensors that belong to a Collector.

Changes in the physical network are directly mapped to the internal data structures of SensorNet. Therefore, queries about network state such as network topology and sensor status do not have to be forwarded to the physical network. This reduces network traffic and energy consumption.

Sensornet also periodically checks whether its resources are being updated. This periodicity is a parameter that the Collector has to inform, as each resource for each particular application has different timing requirements. A sensor, collector or network resource that has not received an update for this period has its state changed to INACTIVE. A subsequent update changes the state to REACTIVATED. A partially functioning sensor, e.g., it should inform temperature and luminosity but only temperature is working, is in MALFUNCTION state. Added resources are in state NEW until they are updated, changing the state to UPDATED. Resources properly working remain forever in state UPDATED. State change messages are sent to message group omcp://notification.messagegroup.osiris/.

4.3.3 VirtualSensorNet

VirtualSensorNet is a module responsible for decoupling sampled data from physical sensors. For example, a temperature monitoring application for datacenters may be interested in obtaining temperature values in front of rack 37. It does not matter how this data is obtained: using sensor of type A or B, calculating the average between the temperature of adjacent racks, etc. VirtualSensors offer this level of abstraction, and VirtualSensorNet is the module that manages VirtualSensors.

There are three types of VirtualSensors: *Link*, *Composite*, and *Blending*. VirtualSensorLink binds to a sensor resource in SensorNet in order to obtain sampled data. If a physical sensor fails and is replaced, VirtualSensorLink may bind to the new sensor and this update is transparent to consuming applications.

VirtualSensorComposite may bind to any number of VirtualSensors of any type and select which of their fields to use. For example, if one VirtualSensorLink has temperature and luminosity values of a mobile device and another VirtualSensorLink has the GPS location this device, we can create a VirtualSensorComposite containing GPS location and temperature. VirtualSensorComposites do not store their sampled data in a database, they link to the original data source; links may be nested.

VirtualSensorBlending allows results of processed sampled data to be published as a sensor reads. For example, if a server room has 30 temperature sensors spread over the area, a VirtualSensorBlending could be used to publish the average temperature of the room as if there existed a sensor to monitor average temperature. This abstraction is very powerful as it can be used to obtain fault tolerance, data conversion, etc.

In order to process data, VirtualSensorBlending binds upon creation to a Function module (see Section 4.3.4) and to any number of VirtualSensors of any type to select which of their fields to use. Periodically (this periodicity is informed upon VirtualSensorBlending creation), the Function module processes data from VirtualSensors and the result is published as a sensor read.

Similar to SensorNet, VirtualSensorNet also broadcasts VirtualSensor state changes to message group omcp://notification.messagegroup.osiris/.

4.3.4 Function and External

Functions are application-specific modules that process input data and return a result. They are used by VirtualSensorBlending, as described in Section 4.3.3. External modules are wildcards that perform any particular action of a monitoring application. The most trivial use of this module is to consume data for visualization. External modules may send messages to other modules, e.g., to retrieve history data, to instruct the creation/deletion of VirtualSensors, etc. We plan on using this module to implement a GUI SensorNet and VirtualSensorNet Manager.

4.4 Evaluation

In this section, we assess the performance of OSIRSIS measuring the end-to-end latency, henceforth called *E2ELatency*, from the moment each sample arrives at the Collector until the sample arrives at a consuming application module. Our experimental setup consists of two machines directly connected over a 10 Gbits cabled link: one hosts OSIRIS modules and the other hosts the communication layer. Collectors were configured to generate synthetic random data and publish them at different rates. There is no need to set up a real WSN to generate data as we just want to assess OSIRIS performance under different workloads. We emphasize that OSIRIS capability to support the implementation of sensor-based monitoring systems has been assessed in the previous work [15].

In our first set of experiments, we create one virtual sensor of type link (VSensorLink) and use one Collector to generate its data; data is generated within intervals of 2 s. For each scenario in this set, we vary the number of messages that are sent in each interval and run 100 experiments of 100 intervals each.

Figure 4.3 depicts in a boxplot our results when one message is sent every 2 s. As can be seen, most E2ELatency values (~75%) are below 32 ms and all but one value are below 50 ms. The highest measured E2ELatency, an exceptional case, is around

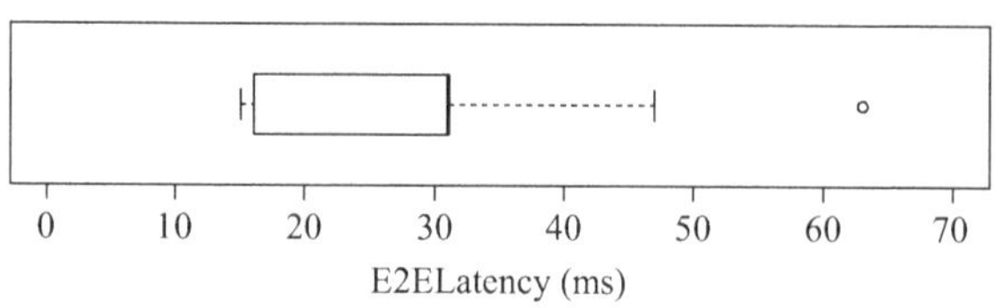

Figure 4.3 E2ELatency for one message every 2 s

62 ms. These results are very good given some commercial WSN products promise 63 ms delay [22].

Figure 4.4 depicts in a barplot the minimum, maximum and average E2ELatency when 10, 100, and 300 message are sent every 2 s. The average bar also plots the 95% confidence interval. In each experiment, the Collector establishes one single connection with the message broker to send all messages. We can see that in this case OSIRIS can barely handle 300 messages per second.

If each Collector establishes one dedicated connection with the message broker to send each message, we can see in Figure 4.5 that OSIRIS can handle up to 10.000 messages per second with maximum E2ELatency of 1 s and average of around 30 ms. We believe that using one connection causes latter sampled data to stall, while multiple parallel connections avoid this bottleneck. This is an important observation for developers using OSIRIS to build their monitoring systems.

In our second set of experiments, we create one virtual sensor of type link (VSensorLink), 100 nested composite virtual sensors (VSensorComposite001 to

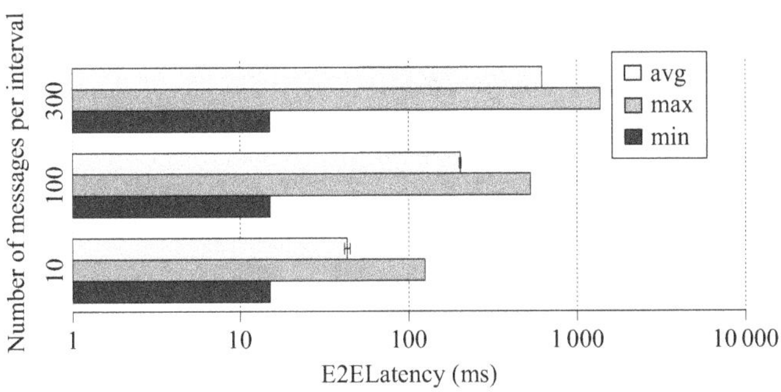

Figure 4.4 One connection per Collector for all messages

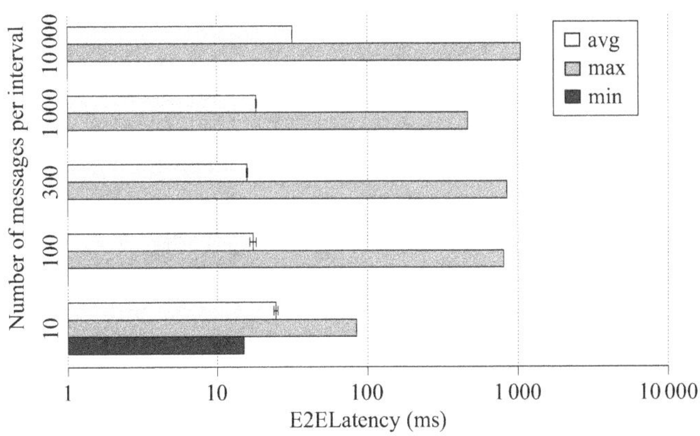

Figure 4.5 One dedicated connection for each message

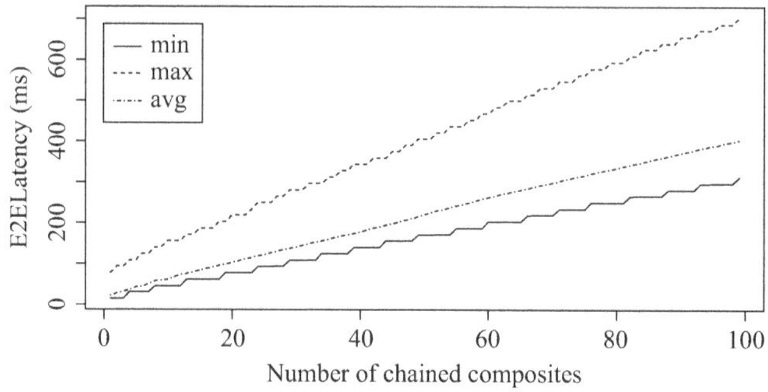

Figure 4.6 Overhead of nested composite virtual sensors

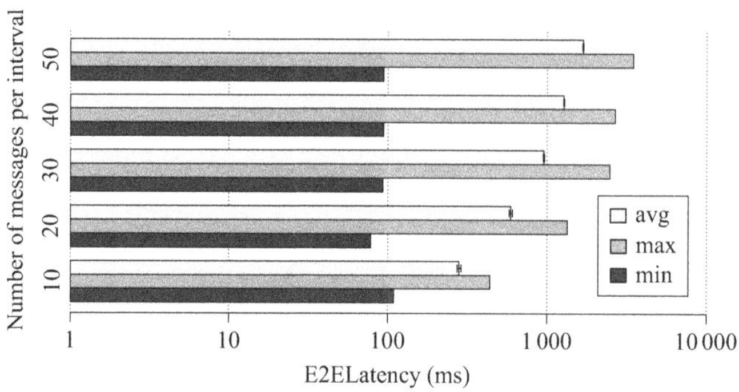

Figure 4.7 Overhead of several Collectors on same machine

VSensorComposite100) and use one Collector to generate data; data is generated with intervals of 2 s. Nested VSensorComposite means that VSensorComposite001 is linked to VSensorLink, VSensorComposite002 is linked to VSensorComposite001, and so on. We send a total of 100 messages and plot the minimum, maximum ad average E2ELatency for each VSensorComposite in Figure 4.6. We can see that the average E2ELatency is close to the minimum and that latency increases linearly with the nesting depth. Therefore, composite virtual sensors scale well but should not be overused so performance is not compromised.

Finally, in our third set of experiments, we create one virtual sensor of type link (VSensorLink) and vary the number of Collectors that generate data. Each collector generates one message every 2 s and each experiment consists of sending 100 messages. Figure 4.7 plots the minimum, maximum, and average E2ELatency for 100 experiments as we vary the number of Collectors that generate data simultaneously. We can see that with 30 Collectors the average E2ELatency already reaches 1 s. From our first set of experiments, we expected to be able to handle 10.000 Collectors since each Collector sends only one message. We believe that this extra overhead results from running all Collectors on the same machine. In future work, we want to run each Collector on a different machine.

4.5 Conclusion

In this work, we proposed the OSIRIS Framework, an infrastructure to build monitoring systems that collect, process, store, and share data to multiple distributed and concurrent applications. OSIRIS offers tools to handle data from different sources in a standardized and flexible way and to manage the sensor network while hiding intrinsic details of how data are obtained. Our tests show that OSIRIS can handle a traffic rate up to 10.000 samples per second, which suffices for large-scale commercial applications of WSN. Multiples sink distributed over a wide area is not a problem since our collector modules work distributed. Source codes and binaries are publicly available on GitHub (https://github.com/labtempo/osiris and https://github.com/labtempo/osiris-binaries).

References

[1] J. Yick, B. Mukherjee, and D. Ghosal. "Wireless sensor network survey." *Computer Networks*, vol. 52, no. 12, pp. 2292–2330, 2008.

[2] M. Suzuki, S. Saruwatari, N. Kurata, and H. Morikawa. "A high-density earthquake monitoring system using wireless sensor networks." In *Proceedings of the 5th International Conference on Embedded Networked Sensor Systems, SenSys '07*, New York, NY, USA, 2007. ACM, pp. 373–374.

[3] B. Wang, X. Guo, Z. Chen, and Z. Shuai. "Application of wireless sensor network in farmland data acquisition system." In J. Zhang, editor, *Applied Informatics and Communication, Communications in Computer and Information Science*, vol. 226, Springer Berlin Heidelberg, 2011, pp. 672–678.

[4] G. Zanatta, G.D. Bottari, R. Guerra, and J.C.B. Leite. "Building a WSN infrastructure with COTS components for the thermal monitoring of datacenters." In *Symposium on Applied Computing, SAC 2014*, Gyeongju, Republic of Korea, pp. 1443–1448, March 24–28, 2014.

[5] R. Piyare, and S. R. Lee. "Towards internet of things (iots): Integration of wireless sensor network to cloud services for data collection and sharing." *CoRR*, abs/1310.2095, 2013.

[6] S. Kelly, N. Suryadevara, and S. Mukhopadhyay. "Towards the implementation of iot for environmental condition monitoring in homes." *Sensors Journal, IEEE*, vol. 13, no. 10, pp. 3846–3853, 2013.

[7] D. Zhong, H. Lv, J. Han, and Q. Wei. "A practical application combining wireless sensor networks and internet of things: Safety management system for tower crane groups." *Sensors*, vol. 14, no. 8, pp. 13794–13814, 2014.

[8] S. Madria, V. Kumar, and R. Dalvi. "Sensor cloud: a cloud of virtual sensors." *Software, IEEE*, vol. 31, no. 2, pp. 70–77, 2014.

[9] H. Durrant-Whyte, and T. Henderson. "Multisensor data fusion." In B. Siciliano and O. Khatib, editors, *Springer Handbook of Robotics*, Springer Berlin Heidelberg, pp. 585–610, 2008.
[10] A. Kansal, S. Nath, J. Liu, and F. Zhao. "SenseWeb: an infrastructure for shared sensing." *IEEE MultiMedia*, vol. 14, no. 4, pp. 8–13, 2007.
[11] S. Kabadayi, A. Pridgen, and C. Julien. "Virtual sensors: abstracting data from physical sensors." In *World of Wireless, Mobile and Multimedia Networks, 2006. WoWMoM 2006. International Symposium on a*, pages 587–592, 2006.
[12] F. Chen, and R. Li. "Sink node placement strategies for wireless sensor networks." *Wireless Personal Communications*, vol. 68, no. 2, pp. 303–319, 2013.
[13] D. Networks. "Wireless sensor networks make it possible to predict precious water supplies." Technical report, Linear Technology Corporation, 1630 McCarthy Blvd. Milpitas, CA.
[14] Vigilent and D. Networks. "Close the loop on energy management at the california franchise tax board." Technical report, Linear Technology Corporation, 1630 McCarthy Blvd. Milpitas, CA.
[15] F. Santos, and R. Guerra. "Osiris framework: construindo sistemas de monitoramento com redes de sensores sem fio para compartilhar dados." In *Anais do 33 Simposio Brasileiro de Redes de Computadores e Sistemas Distribuidos*, Vitoria/ES, Brazil, 2015.
[16] J. Russell, and R. Cohn. *Rabbitmq*. "World Chess Championship 2012." Book on Demand, 2012.
[17] R.T. Fielding. "*Architectural styles and the design of network-based software architectures*." PhD thesis, 2000. AAI9980887.
[18] R. Fielding, J. Gettys, J. Mogul, *et al.* Hypertext transfer protocol – http/1.1, 1999.
[19] Pivotal. "Rabbitmq hits one million messages per second on google compute engine." https://blog.pivotal.io/pivotal/products/rabbitmq-hits-one-million-messages-per-second-on-google-compute-engine, 2015. Accessed: 2015-11-25.
[20] S. Vinoski. "Advanced message queuing protocol." *IEEE Internet Computing*, vol. 10, no. 6, pp. 87–89, 2006.
[21] Pivotal. "AMQP 0-9-1 model explained." https://www.rabbitmq.com/tutorials/amqp-concepts.html, 2014. Accessed: 2014-12-04.
[22] B.E. Corp. "Zero defect standard achieved using wireless sensor solution on industrial winder." Technical report, Banner Engineering Corp, 9714 Tenth Ave. No., Minneapolis, MN, USA.

Chapter 5

Modeling and tracing events in RFID-enabled supply chains

Cong-cong Xing[1], *Jun Huang*[2] *and Shui Yu*[3]

Abstract

With cloned tag attacks being a serious problem in radio frequency identification (RFID) supply chains, tracing event records and inspecting them are one of the notably effective means to defense such attacks. Unfortunately, current event formulations are unable to characterize the variety and complexness of the events in real-world RFID supply chains. This chapter identifies the problem of the existing tag event models, reviews the RFID supply chain architecture, proposes a new tag event model, and devises the ensuing event-tracing algorithm. It is our belief that the newly proposed event model together with the event-tracing algorithm based on this model furnish a foundation for future more sophisticated event-record-based clone detection techniques.

5.1 Introduction

The architecture of the Internet of Things (IoT) can be roughly sketched as consisting of a bottom sensor layer, a middle network layer, and a top application layer. As one of the primary information-acquiring means at the bottom layer of the IoT, radio frequency identification (RFID) tags have found increasingly widespread applications in various business areas, with the expectation that the use of RFID tags will eventually replace the existing bar codes in all business areas. While the RFID technology offers numerous exciting new benefits, it also brings in new challenges. The RFID tag of each product contains an electronic product code (EPC) that is used to uniquely identify this product. However, the EPC contents of an RFID tag can be easily cloned by some malicious attackers, and the cloned tags may be attached to some

[1]Department of Mathematics/Computer Science, Nicholls State University
[2]Institute of Electronic Information and Networking, Chongqing University of Posts and Telecommunications
[3]School of Information Technology, Deakin University

counterfeit products which then can be injected into business RFID supply chains, transit along the supply chains subsequently, and are eventually sold to customers.

As such, techniques, methods, and strategies on how to prevent counterfeit RFID/EPC tags from being made as well as on how to detect counterfeit tags (once they are in the supply chain) have been extensively studied in the literature (e.g., see References 1–3). Between these two approaches, it has been realized that there are essentially no feasible and effective ways to completely prevent an RFID tag from being cloned. Hence, much efforts have been devoted to the cloned tag detection techniques. One notable method of detecting cloned RFID tags is to check the history record of events of the suspected tag against some predefined criteria. To use this approach to detect cloned tags, one must access the Discovery Service of the EPC-global network to obtain a set of desired events first, which immediately raises the issue of modeling events and tracing events in RFID supply chains.

Currently, RFID tag-related events are typically treated as *simple* events in the literature in the sense that there is no need for any tag changes, and a tag remains isolated during its entire course of transition in supply chains. Unfortunately, this is not the case in real-world commercial RFID supply chains where events are more complicated. For instance, a box of goods may be split into sub-boxes of goods at a distribution center, and several (different types of) products may be bundled together for sale at a wholesale store. Both activities involve tag changes in their courses. Toward resolving this problem, we in this chapter model RFID tag events by taking the variety of events into consideration, and address the ensuing event history record issue by devising a corresponding event-tracing algorithm.

The remainder of this chapter is organized as follows. Section 5.2 reviews the related work in the literature, Section 5.3 describes the structural elements in RFID-enabled supply chain systems. Section 5.4 presents a model of the RFID tag event that encapsulates an array of different types of events, which is followed in Section 5.5 by a tag event-tracing algorithm together with some illustrating examples. Section 5.6 concludes this chapter.

5.2 Background and related work

The security issues associated with RFID supply chains are wide-spread and far-reaching [4–10]. With regard to the strategies for defending cloned tag attacks, as mentioned earlier in Section 5.1, there exist two different approaches: prevention and detection.

The primary means to prevent cloned RFID tags from being made is to encrypt the data contained in RFID tags via public/private keys to make it difficult to be revealed and cloned, and many such mechanisms have been proposed. Lehtonen *et al.* [11] proposed to use the memory of RFID tag to save a secret key which will be updated every time the tag goes through a successful reading. Abawajy [12] suggested a mechanism to enhance RFID tags' resistance against being cloned by protecting RFID tags' identifications and recognitions. As an emerging encryption technique, physically unclonable function devices [13,14] use the "finger print" that is unique and intrinsic to each piece of semiconductor hardware to secure the secret key. Closely related to

RFID tag data encryption is the RFID communication authentication. Two lightweight RFID authentication protocols are proposed in References 15 and 16, respectively, where the first protocol allows tags and readers/writers to authenticate to each other, and the second protocol only requires Hash or XOR operations be performed in tag memory. Shen *et al.* [17] presented a novel anonymous RFID authentication protocol which not only provides authentications but disguises RFID tags against attackers by giving them fake names, as well.

Unfortunately, encryption of data and the subsequent authentication require both complex algorithms/computations and communication overheads between RFID tag memory and the EPC reader/writer, and thus generally cannot be sustained by the typically simple, lightweight, resource-limited RFID tags (e.g., the standard EPC C1G2 tags). Moreover, a cloned RFID tag, although not being the original tag, is a perfect duplication of the original tag possessing not only an identical EPC content but also a legitimate private key, and therefore is able to pass the identity authentication in the same way as an authentic RFID tag would. Since there is essentially no effective way to prevent cloned RFID tags from being made [11], extensive studies have been conducted on the detection of cloned tags after they are injected into RFID supply chains.

Mirowski and Hartnett [18] investigated a method that detects the change of the RFID tag ownership by checking the combination of reader/writer operations, tag and reader/writer identifications, and the time stamps of tag events. Kerschbaum and Oertel [7] proposed a machine-learning technique that detects clone existence suggesting tag anomaly by pattern matching. By noting the cloned tags and authentic tags always respond at the same time to the message sent by the reader/writer, and these simultaneous responses will inevitably collide in an unreconciled fashion, Bu *et al.* [19] suggested that cloned tags can be uncovered by such unreconciled collisions. A batch clone detection mechanism was discussed in Reference 20 where each supply chain partner generates a random number when it processes a batch of RFID products. This generated random number is sent to a central server and the server detects the existence of cloned tags by checking the consistency of the generated numbers. Another method of detecting cloned RFID tags is to check the history record of events of the suspected tag against some predefined criteria [1,3]. Basically, if all criteria are passed, then the suspected tag is a genuine tag; otherwise, it has a strong probability of being a cloned tag.

The topic of this chapter is the modeling and tracing of tag events, which serve as the foundation for event history record-based clone detection techniques.

5.3 The RFID-enabled supply chain system

As a preparation for the modeling of the RFID supply chains in the next section, here, we give an overview of the RFID supply chain system.

5.3.1 System architecture

The essential architecture of an RFID supply chain system is illustrated in Figure 5.1. Every node in the supply chain system/network represents some supply chain participant. As such, each node in the supply chain system serves as a node in the larger

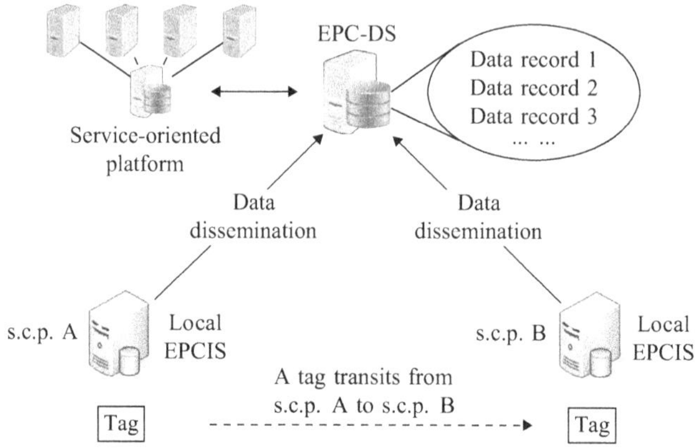

Figure 5.1 The architecture of RFID-enabled supply chains (s.c.p. = supply chain participant)

EPC network, and thus the transition of a product (with an EPC tag) in the supply chain network naturally gives rise to the flow of the product's information on the larger EPC network. Each supply chain partner is equipped with multiple RFID readers/writers, has a local EPC Information Service (EPCIS) database, and can send product-related event information (such as product's EPC and the address of the local EPCIS) to the remote RFID/EPC Discovery Service (DS). When an EPC tag is read by a reader at the site of a supply chain partner, a corresponding event will be created and saved in the local EPCIS database. Since a tag will be read many times at many different sites, various events associated with one tag will thus be stored, in a distributed fashion, at different local EPCIS databases of supply chain participants. Each supply chain participant (or local EPCIS database) may choose to share, via RFID infrastructure (EPCglobal network), any event information with other supply chain participants.

As a service-oriented platform, the DS can be consulted by a third-party program (e.g., a counterfeit product detection program) in its process of tracing all events related to this product, and in the program's subsequent analysis and decision-making, once a product is sold at a retailer which triggers the execution of the program. Also, a third-party program may be allowed to access the local EPCISs to collect required data and analyze relevant events via authentication and/or access control mechanisms [21,22], which is useful in terms of the optimization of the system.

Throughout this chapter, we make the following assumptions/stipulations for the supply chain system.

1. A product and its EPC tag are regarded as one single item that cannot be separated.
2. Each tag has a limited amount of memory with basic functionalities only. There is no communication authentication between a tag and the tag reader/writer, which is consistent with the widespread use of the unpowered tags, especially EPC C1G2 tags, in the real-world RFID supply chains.

3. The identification of a tag cannot be altered. But the contents of a tag's memory can be read and (re)written.
4. An event will be created and saved in the local EPCIS database whenever a tag is being read or written. The information contained in such an event should include the EPC of the product, the time when this event occurs, the location of the local supply chain participant, and the nature of the business transaction (e.g., shipping, receiving, and inventory).
5. Two special events (beginning event and ending event) will be created, respectively, when a product enters the supply chain and exits the supply chain. The former characterizes the activity when an EPC tag is assigned to a product at a manufacturer, and the latter characterizes the activity of a product being sold at a retailer.
6. Supply chain participants generally agree to share the event data with one another.
7. Any supply chain participant only knows its direct business partners, and may choose to leave or join the supply chain at any time.

5.3.2 The discovery service mechanism

In RFID-enabled supply chains, only supply chain participants (companies) that have processed an EPC tag are allowed to query information related to this specific tag by sending requests to the DS. In terms of the working mechanism of the existing DS in real-world supply chains, there are the following four models: (1) catalog service model, (2) query delay model, (3) data consolidation model, and (4) secure data model, as shown in Figure 5.2. We briefly explain each of them below.

- *Catalog service model:* The DS will return the relevant web addresses (URL) to the querying company when the DS receives the EPC-searching request from the querying company. By using the obtained URL, the querying company will contact the target company directly and acquire the desired results.
- *Query delay model:* The DS does not return any URL to the querying company when it receives such a querying request; instead, it redirects the query to the data-sending company. The data-sending company will return the desired results directly to the data-querying company.
- *Data consolidation model:* The DS works in a similar fashion to that of the query delay model, except that it receives the results returned by the data-sending company, consolidates it, and sends it to the data-querying company.
- *Secure data model:* Secure data model [23] is an improved version of the catalog service model in the sense that both access to the DS database and access to the data-sending company are securely controlled for the protection of data. Currently, secure data model is the only DS mechanism that considers the security of data and protects the privacy of business companies.

5.3.3 Access controls of the secure Data DS

For the sake of privacy of product information, some kind of control over the access to the product information must be implemented in the mechanism of the secure data

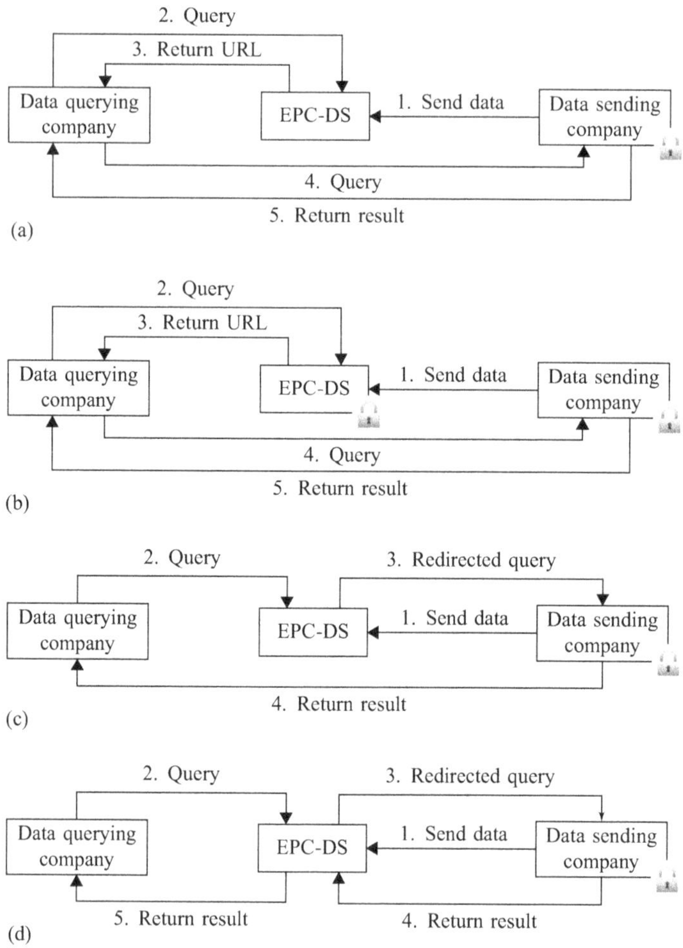

Figure 5.2 Discovery service models. (a) Catalog service model; (b) secure data model; (c) query delay model; (d) data consolidation model

model of DS. When a data-sending company releases an event, the related access control method to this event will be released together with the event. Some basic strategies regarding business partner identity authentication and product information access-right authorization in RFID supply chains are proposed by Kywe *et al.* [22], where the identity authentication can be handled by the X.509 certification, and the access-right authorization can be controlled at three levels: public, protected, and private. If the access control for a product is labeled as "public," then every individual is allowed to receive the information about this product from the DS; if it is labeled as "protected," then only companies which are related to this product in terms of business are allowed to receive the information about this product; and if it is labeled as "private," then only the owner itself can access the information about this product, nobody else can.

A more sophisticated access control mechanism – attribute-based access control (ABAC) – is proposed in References 24 and 25. Specifically, in ABAC, a product can be characterized by a combination of subject attributes, object attributes, and visibility attributes.

- *Subject attribute:* Subject attributes refer to properties that are related to the supply chain participant (company) itself, such as the company's name, ID, role, and location.
- *Object attribute:* Object attributes refer to properties that are related to the EPC events associated with the product, such as the EPC itself, the time and location of the event, and the business itinerary of the product.
- *Visibility attribute:* Visibility attributes refer to the relative business relationship among supply chain participants, and they can be one of the following values: "upper," "lower," and "all." For example, consider a supply chain that includes raw material suppliers, manufacturers, distributors, wholesalers, and retailers. Then, for a distributor, raw material suppliers and manufacturers will be considered its "upper" partners; wholesalers and retailers will be considered its "lower" partners; and all participants will be considered its "all" partners.

By combining the above attributes, a data-sending company is able to stipulate the access control rubrics for a product when disseminating an EPC event associated with this product. For instance, it can be controlled that some events can only be viewed by manufacturers, that some companies can only access events that have occurred during a specific time period, and that some events can only be viewed by their "upper" or "lower" partners.

5.4 Modeling of the system

We now address the issue of modeling the RFID supply chains.

5.4.1 Events

The RFID-enabled supply chains described in the previous section can be intuitively and initially modeled as a directed graph such as the one shown in Figure 5.3, where each node represents a supply chain participant (manufacturer, distribution center, distributor, or retailer) labeled with EPC tags being processed at that node, and each edge represents the transition (regular, split, combined, or recall, see explanation below) of a tag from one supply chain participant to another. For example, the shipment of a product with tag T_1 from manufacturer P_1 to distribution center P_3 is a regular transition, while the shipment of a product with tag T_3 from distributor P_{10} to retailer P_{11} is a combined transition.

An event takes place when a supply chain participant receives a product, ships a product, or processes a product. By the studies on EPC tags and supply chains (e.g., see References 26 and 27), events occurring on supply chains can be essentially classified into the following categories:

- *Regular event*: A product/tag transits from one supply chain participant to another without any changes. Regular events are marked by "r" in our initial model of supply chains. For example, the edge $P_1 \xrightarrow{r} P_3$ in Figure 5.3 represents a regular event.
- *Split event*: A product/tag transits from one supply chain participant to another by splitting its original EPC tag into a set of new tags. For example, a case of beer with tag T_α containing four six-packs of beer may be split into four individual six-packs with tags T_a, T_b, T_c, and T_d, respectively, which then can be sent to four different supply chain participants. Split events are marked by "s" in our initial model of supply chains. In Figure 5.3, the edges $P_4 \xrightarrow{s} P_5$, $P_4 \xrightarrow{s} P_6$, and $P_4 \xrightarrow{s} P_7$ are all split events.
- *Combined event*: A set of products with different tags and from different supply chain partners may transit to one supply chain participant to form a new product with a new EPC tag. For example, a bottle of beer with tag T_x and a rose with tag T_y may be combined together to form a new gift product with a new tag T_z. Combined events are marked by "c" in our initial model of supply chains. In Figure 5.3, the edges $P_{10} \xrightarrow{c} P_{11}$, and $P_9 \xrightarrow{c} P_{11}$ are all combined events.
- *Recall event*: A product/tag transits back a manufacturer due to the manufacturer's recall on products with defects. Recall events are marked by "rc" in our initial model of supply chains. For example, in Figure 5.3, the edge $P_2 \xrightarrow{rc} P_1$ is a recall event.

Toward defining events formally, we need to take a closer look at the nature of events. Note that a split event actually occurs at a supply chain participant after some product has been received, and then the split products will be sent out to other supply

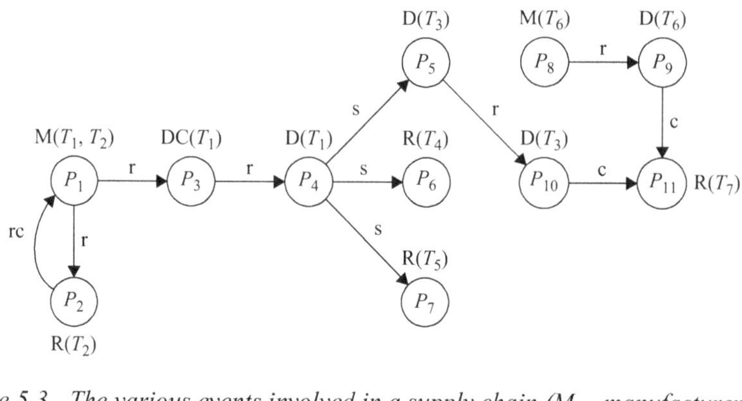

Figure 5.3 The various events involved in a supply chain (M = manufacturer, DC = distribution center, D = distributer, R = retailer, r = regular transition, s = split transition, c = combined transition, rc = recall transition)

chain participants. For example, the edge $P_4 \xrightarrow{s} P_5$ in Figure 5.3 in fact involves the sequence of the following three activities: (1) distributer P_4 splits a product with tag T_1 into three products with tags T_3, T_4, T_5, respectively, (2) distributor P_4 ships the product with tag T_3 to distributor P_5, and (3) the distributor T_5 receives it. Clearly, the split activity occurs within the boundary of distributor P_4 *before* the product with tag T_3 is shipped to P_5. The combined events can be characterized in the same fashion. As such, we refine the types of events as follows:

- *Input event*: This event describes the fact that a supply chain participant receives a product from another supply chain participant. There is no product tag change in this event since the product is simply (shipped and) received.
- *Output event*: This event describes the fact that a supply chain participant ships a product to another supply chain participant. There is no product tag change in this event since the product is simply shipped (and received).
- *Split event*: This event describes the fact that a supply chain participant divides one product into several different products. There *is* a product tag change in this event.
- *Combined event*: This event describes the fact that a supply chain participant combines several different products into one product. There *is* a product tag change in this event. (This event can be viewed as the reversal of the split events.)
- *Recall event*: This event describes the fact that a supply chain participant ships a product with defects back to the manufacturer. There is no product tag change in this event for the obvious reason.

Based on the refined types of events, we can formally define an event in an RFID-enabled supply chain as follows. Let $\mathbb{EPC}$ be the set of all EPC tags, $\mathbb{ET} = \{i, o, s, c, rc\}$ be the set of event types (representing input, output, split, combined, recall events, respectively), $\mathbb{L}$ be the set of locations of supply chain participants, and $\mathbb{T}$ be the set of time stamps. Then an event is a four-tuple

$$(tp, et, l, t) \in (\mathbb{EPC} \times \mathbb{EPC}) \times \mathbb{ET} \times \mathbb{L} \times \mathbb{T}$$

where

- $tp = \langle fEPC, nEPC \rangle \in \mathbb{EPC} \times \mathbb{EPC}$ is a tag pair, with $fEPC$ being the initial tag and $nEPC$ being the end tag,
- $et \in \mathbb{ET}$ is the event type,
- $l \in \mathbb{L}$ is the location where the event occurs,
- $t \in \mathbb{T}$ is the time stamp when the event occurs,

such that $fEPC = nEPC$ if $et = i, o$, or rc, and $fEPC \neq nEPC$ otherwise.

Note that the last condition in the above definition ensures that there would be no tag changes in input, output, and recall events, and there would be some tag changes for split and combined events, as we observed earlier. For example, with reference to Figure 5.3, the event $(\langle T_1, T_1 \rangle, o, P_1, t_2)$ represents the fact that a product with tag T_1 is shipped out of the manufacturer P_1 at time t_2, and the receiver receives the same product with tag T_1; the event $(\langle T_1, T_3 \rangle, s, P_4, t_8)$ represents the fact that a product with tag T_3 is produced by splitting a product with tag T_1 at distributor P_4 at

Table 5.1 All events that are associated with the supply chain in Figure 5.3

Supply chain participant	Events
P_1	$(\langle T_1, T_1\rangle, o, P_1, t_2), (\langle T_2, T_2\rangle, o, P_1, t_1)$
P_2	$(\langle T_2, T_2\rangle, rc, P_2, t_4), (\langle T_2, T_2\rangle, i, P_2, t_3)$
P_3	$(\langle T_1, T_1\rangle, o, P_3, t_6), (\langle T_1, T_1\rangle, i, P_3, t_5)$
P_4	$(\langle T_3, T_3\rangle, o, P_4, t_{11}), (\langle T_4, T_4\rangle, o, P_4, t_{10}), (\langle T_5, T_5\rangle, o, P_4, t_9),$ $(\langle T_1, T_3\rangle, s, P_4, t_8), (\langle T_1, T_4\rangle, s, P_4, t_8), (\langle T_1, T_5\rangle, s, P_4, t_8),$ $(\langle T_1, T_1\rangle, i, P_4, t_7)$
P_5	$(\langle T_3, T_3\rangle, o, P_5, t_{13}), (\langle T_3, T_3\rangle, i, P_5, t_{12})$
P_6	$(\langle T_4, T_4\rangle, o, P_6, t_{15}), (\langle T_4, T_4\rangle, i, P_6, t_{14})$
P_7	$(\langle T_5, T_5\rangle, o, P_7, t_{17}), (\langle T_5, T_5\rangle, i, P_7, t_{16})$
P_8	$(\langle T_6, T_6\rangle, o, P_8, t_{24})$
P_9	$(\langle T_6, T_6\rangle, o, P_9, t_{26}), (\langle T_6, T_6\rangle, i, P_9, t_{25})$
P_{10}	$(\langle T_3, T_3\rangle, o, P_{10}, t_{19}), (\langle T_3, T_3\rangle, i, P_{10}, t_{18})$
P_{11}	$(\langle T_7, T_7\rangle, o, P_{11}, t_{23}), (\langle T_3, T_7\rangle, c, P_{11}, t_{22}), (\langle T_6, T_7\rangle, c, P_{11}, t_{22}),$ $(\langle T_3, T_3\rangle, i, P_{11}, t_{21}), (\langle T_6, T_6\rangle, i, P_{11}, t_{20})$

time t_8; the event $(\langle T_3, T_7\rangle, c, P_{11}, t_{22})$ represents the fact that a product with tag T_7 is assembled at retailer P_{11} by combining a product with tag T_3 and other products, at time t_{22}; and the event $(\langle T_5, T_5\rangle, i, P_7, t_{16})$ represents the fact that a product with tag T_5 (is shipped by a sender first, and then) enters the retailer P_7 at time t_{16}.

For the sake of convenience and the study of event-tracing algorithm later in this chapter, we enumerate all events associated with Figure 5.3 in Table 5.1, and index them by the supply chain participants. Note that in Table 5.1, the subscript of time in events does not suggest any order for associated time stamps. That is, $i < j$ does *not* necessarily imply that $t_i < t_j$ (or t_i is earlier than t_j). Those subscripts are used purely to indicate different time stamps.

Note, however, that there is not a one-to-one correspondence between the edges in Figure 5.3 and the events. For instance, the edge $P_4 \xrightarrow{s} P_5$ in fact involves three different events: a split event, an output event, and an input event, as mentioned earlier. For this reason, we refine the initial supply chain graph model and formally define it as follows. An RFID-enabled supply chain is an *attributed* directed graph

$$(V, E, A, src, tgt, att)$$

where

- V is the set of vertices (supply chain participants).
- E is the set of edges.
- A is the set of attributes (events).
- *src* is the source function from E to V.
- *tgt* is the target function from E to V.
- *att:* $E \rightarrow \mathcal{P}(A)$ is the attribute function from E to $\mathcal{P}(A)$, where $\mathcal{P}(A)$ denotes the power set of A. That is, *att* maps an edge to a set of events that are associated with this edge.

By this definition, Figure 5.3 can be precisely specified or reformulated by adding the attributes of its edges. For example, the attribute of the edge $e = P_4 \xrightarrow{s} P_5$ is $att\,(e) = \{(\langle T_1, T_3\rangle, s, P_4, t_8),\ (\langle T_3, T_3\rangle, o, P_4, t_{11}), (\langle T_3, T_3\rangle, i, P_5, t_{12})\}$, which models the intended meaning of the supply chain exactly.

5.4.2 Event dissemination

DS is a centralized database which keeps the record of all events that have occurred at some point in a supply chain, and all supply chain participants' information such as their company addresses, phone numbers, website addresses, product data, and partnership with other companies. The primary purpose of DS is to provide services to supply chain participants upon their query requests, but the information held at the DS is private and thus is not available to the general public.

Local EPCISs are the main resources from which DS acquires its data. Critical pieces of information stored in local EPCISs will be singled out and disseminated to the DS, so that DS, as a lightweight catalog service system, may allow customers to locate all information linked to a target product [7–10]. As such, the minimum amount of data in a DS-record must include a product identifier (i.e. EPC), a pointer (referring to the information resources about the product), and a time stamp (recording the time when the record is created).

Any supply chain participant can either register event information to DS, or request the DS to search some information about a particular supply chain partner. Particularly, a supply chain participant registers an input event to the DS when it receives a product, a split event when it decomposes a product into several other products, a combined event when it assembles a product by piecing several products together, and an output event when it ships a product to another supply chain partner.

5.5 Tracing events

Given the model of events developed in the previous section, we now turn our attention to the event tracing,

5.5.1 The algorithm

Given a tag EPC *index*, the following recursive procedure *Trace(index)* finds all events associated with *index*, and stores them in a global structure S_T through updating the contents of S_T repeatedly.

Algorithm: *Trace(index)*
Input: A tag EPC *index*.
Output: The set S_T of all events associated with *index*.

Step 1. $S_T = \emptyset$.
Step 2. Find all events E in DS such that $E.TP.nEPC = index$, and store all these events in set S. (*TP* is the tag pair component of an event and *nEPC* is the second component of a tag pair.)

Step 3. Order all events in S, from early to late, in terms of their time component.
Step 4. Search all events in S, looking for events E with $E.TP.fEPC \neq E.TP.nEPC$.

- If no such events can be found, then there are two cases to deal with.
  - If $S = \emptyset$, then terminate the program.
  - Otherwise, all events in S are regular events. In this case, copy all events in S in the existing order to the front of S_T, purge all events from S (i.e., $S = \emptyset$), and terminate the program.

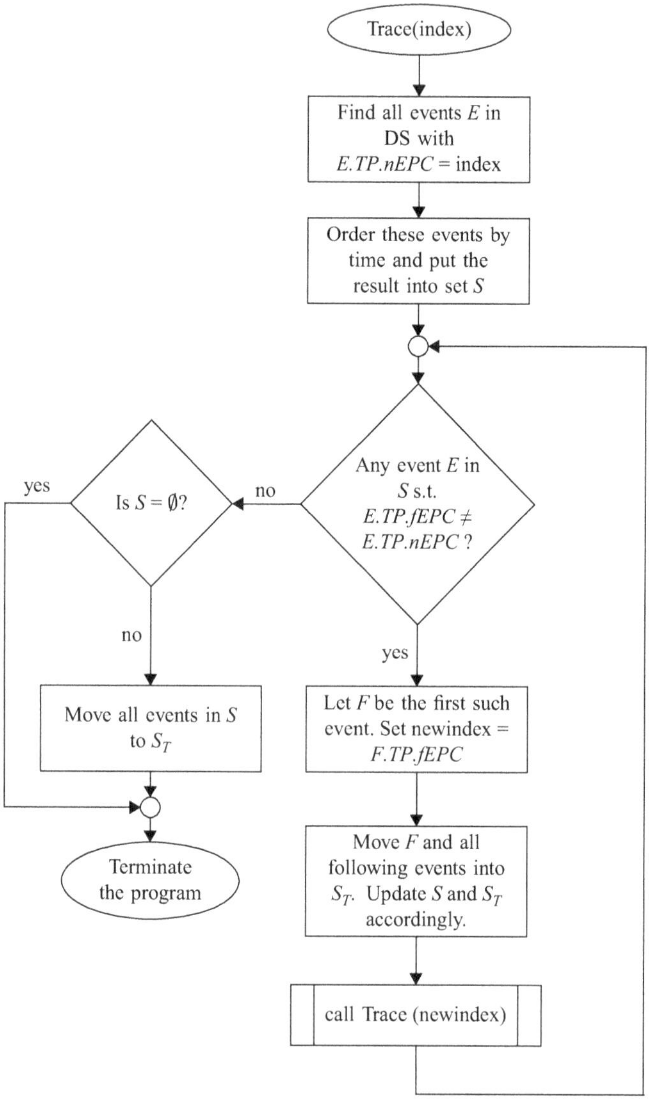

Figure 5.4 Flow chart of the event-tracing algorithm

- Otherwise, proceed to step 5.

Step 5. Let E be the first event in S with $E.TP.fEPC \neq E.TP.nEPC$, and $newindex = E.TP.fEPC$. Copy E and all events after E, in existing order, to the front of S_T, and purge them from S. Update S and S_T accordingly.
Step 6. Call *Trace*(*newindex*) recursively.
Step 7. Go back to step 4 to continue process the remaining events in S.

The flow chart of the above algorithm is depicted in Figure 5.4, which gives a visual explication of the working mechanism of the algorithm.

5.5.2 Event-tracing examples

To see clearly how the event-tracing algorithm works, we present two examples here with reference to Figure 5.3.

Example 1. As the first example, we illustrate the event tracing for the tag T_5. The event-tracing action, $Trace(T_5)$, is triggered when the product with tag T_5 is sold at retailer P_7. Detailed results at stages of the execution of $Trace(T_5)$ are shown in Table 5.2. Basically, the call $Trace(T_5)$ updates S and S_T, and issues a recursive call $Trace(T_1)$. The call $Trace(T_1)$ updates its local S and the global S_T, and returns to its caller $Trace(T_5)$. The continued execution of $Trace(T_5)$ terminates immediately since its local S is empty. The final contents of S_T resulted in by the call $Trace(T_5)$ is: $\{(\langle T_1, T_1\rangle, o, P_1, t_2), (\langle T_1, T_1\rangle, i, P_3, t_5), (\langle T_1, T_1\rangle, o, P_3, t_6), (\langle T_1, T_1\rangle, i, P_4, t_7), (\langle T_1, T_5\rangle, s, P_4, t_8), (\langle T_5, T_5\rangle, o, P_4, t_9), (\langle T_5, T_5\rangle, i, P_7, t_{16}), (\langle T_5, T_5\rangle, o, P_7, t_{17})\}$.

A more comprehensive instance of tag event tracing is given in the next example.

Table 5.2 Computation of Example 1

$Trace(T_5)$	
initial S	$\{(\langle T_1, T_5\rangle, s, P_4, t_8), (\langle T_5, T_5\rangle, o, P_4, t_9), (\langle T_5, T_5\rangle, i, P_7, t_{16}), (\langle T_5, T_5\rangle, o, P_7, t_{17})\}$
initial S_T	$\emptyset$
updated S	$\emptyset$
updated S_T	$\{(\langle T_1, T_5\rangle, s, P_4, t_8), (\langle T_5, T_5\rangle, o, P_4, t_9), (\langle T_5, T_5\rangle, i, P_7, t_{16}), (\langle T_5, T_5\rangle, o, P_7, t_{17})\} \cup$ current S_T
recursive call	$Trace(T_1)$
$Trace(T_1)$	
initial S	$\{(\langle T_1, T_1\rangle, o, P_1, t_2), (\langle T_1, T_1\rangle, i, P_3, t_5), (\langle T_1, T_1\rangle, o, P_3, t_6), (\langle T_1, T_1\rangle, i, P_4, t_7)\}$
initial S_T	$\{(\langle T_1, T_5\rangle, s, P_4, t_8), (\langle T_5, T_5\rangle, o, P_4, t_9), (\langle T_5, T_5\rangle, i, P_7, t_{16}), (\langle T_5, T_5\rangle, o, P_7, t_{17})\}$
updated S	$\emptyset$
updated S_T	$\{(\langle T_1, T_1\rangle, o, P_1, t_2), (\langle T_1, T_1\rangle, i, P_3, t_5), (\langle T_1, T_1\rangle, o, P_3, t_6), (\langle T_1, T_1\rangle, i, P_4, t_7)\} \cup$ current S_T
recursive call	No recursive call. $Trace(T_1)$ terminates and returns.

Table 5.3 Computation of Example 2

Trace(T_7)	
initial S	$\{(\langle T_6, T_7\rangle, c, P_{11}, t_{22}), (\langle T_3, T_7\rangle, c, P_{11}, t_{22}), (\langle T_7, T_7\rangle, o, P_{11}, t_{23})\}$
initial S_T	$\emptyset$
updated S	$\{(\langle T_6, T_7\rangle, c, P_{11}, t_{22})\} \mid \emptyset$
updated S_T	$\{(\langle T_3, T_7\rangle, c, P_{11}, t_{22}), (\langle T_7, T_7\rangle, o, P_{11}, t_{23})\} \mid$ $\{(\langle T_6, T_7\rangle, c, P_{11}, t_{22})\} \cup$ current S_T
recursive call	*Trace*(T_3) | *Trace*(T_6)
Trace(T_3)	
initial S	$\{(\langle T_1, T_3\rangle, s, P_4, t_8), (\langle T_3, T_3\rangle, o, P_4, t_{11}), (\langle T_3, T_3\rangle, i, P_5, t_{12}), (\langle T_3, T_3\rangle, o, P_5, t_{13}),$ $(\langle T_3, T_3\rangle, i, P_{10}, t_{18}), (\langle T_3, T_3\rangle, o, P_{10}, t_{19}), (\langle T_3, T_3\rangle, i, P_{11}, t_{21})\}$
initial S_T	$\{(\langle T_3, T_7\rangle, c, P_{11}, t_{22}), (\langle T_7, T_7\rangle, o, P_{11}, t_{23})\}$
updated S	$\emptyset$
updated S_T	$\{(\langle T_1, T_3\rangle, s, P_4, t_8), (\langle T_3, T_3\rangle, o, P_4, t_{11}), (\langle T_3, T_3\rangle, i, P_5, t_{12}), (\langle T_3, T_3\rangle, o, P_5, t_{13}),$ $(\langle T_3, T_3\rangle, i, P_{10}, t_{18}), (\langle T_3, T_3\rangle, o, P_{10}, t_{19}), (\langle T_3, T_3\rangle, i, P_{11}, t_{21})\} \cup$ current S_T
recursive call	*Trace*(T_1)
Trace(T_1)	
initial S	$\{(\langle T_1, T_1\rangle, o, P_1, t_2), (\langle T_1, T_1\rangle, i, P_3, t_5), (\langle T_1, T_1\rangle, o, P_3, t_6), (\langle T_1, T_1\rangle, i, P_4, t_7)\}$
initial S_T	$\{(\langle T_1, T_3\rangle, s, P_4, t_8), (\langle T_3, T_3\rangle, o, P_4, t_{11}), (\langle T_3, T_3\rangle, i, P_5, t_{12}), (\langle T_3, T_3\rangle, o, P_5, t_{13}),$ $(\langle T_3, T_3\rangle, i, P_{10}, t_{18}), (\langle T_3, T_3\rangle, o, P_{10}, t_{19}), (\langle T_3, T_3\rangle, i, P_{11}, t_{21})\} \cup$ $\{(\langle T_3, T_7\rangle, c, P_{11}, t_{22}), (\langle T_7, T_7\rangle, o, P_{11}, t_{23})\}$
updated S	$\emptyset$
updated S_T	$\{(\langle T_1, T_1\rangle, o, P_1, t_2), (\langle T_1, T_1\rangle, i, P_3, t_5), (\langle T_1, T_1\rangle, o, P_3, t_6),$ $(\langle T_1, T_1\rangle, i, P_4, t_7)\} \cup$ current S_T
recursive call	No recursive call. *Trace*(T_1) terminates and returns.
Trace(T_6)	
initial S	$\{(\langle T_6, T_6\rangle, o, P_8, t_{24}), (\langle T_6, T_6\rangle, i, P_9, t_{25}), (\langle T_6, T_6\rangle, o, P_9, t_{26})\}$
initial S_T	$\{(\langle T_6, T_7\rangle, c, P_{11}, t_{22})\} \cup$ $\{(\langle T_1, T_1\rangle, o, P_1, t_2), (\langle T_1, T_1\rangle, i, P_3, t_5), (\langle T_1, T_1\rangle, o, P_3, t_6), (\langle T_1, T_1\rangle, i, P_4, t_7)\} \cup$ $\{(\langle T_1, T_3\rangle, s, P_4, t_8), (\langle T_3, T_3\rangle, o, P_4, t_{11}), (\langle T_3, T_3\rangle, i, P_5, t_{12}), (\langle T_3, T_3\rangle, o, P_5, t_{13}),$ $(\langle T_3, T_3\rangle, i, P_{10}, t_{18}), (\langle T_3, T_3\rangle, o, P_{10}, t_{19}), (\langle T_3, T_3\rangle, i, P_{11}, t_{21})\} \cup$ $\{(\langle T_3, T_7\rangle, c, P_{11}, t_{22}), (\langle T_7, T_7\rangle, o, P_{11}, t_{23})\}$
updated S	$\emptyset$
updated S_T	$\{(\langle T_6, T_6\rangle, o, P_8, t_{24}), (\langle T_6, T_6\rangle, i, P_9, t_{25}), (\langle T_6, T_6\rangle, o, P_9, t_{26})\} \cup$ current S_T
recursive call	No recursive call. *Trace*(T_6) terminates and returns.

Example 2. This example shows how the result of tracing the tag T_7 is obtained, and is illustrated in Table 5.3. The initial call *Trace*(T_7) yields the top portion of the table and issues a recursive call *Trace*(T_3). The call *Trace*(T_3) generates the second portion of the table and issues another recursive call *Trace*(T_1). The call *Trace*(T_1) produces the third portion of the table and terminates/returns. Note that at this point, the termination of *Trace*(T_1) sends control back to its caller *Trace*(T_3). The continued execution of *Trace*(T_3) immediately terminates itself and sends control back to its caller *Trace*(T_7).

Trace(T_7) now goes into the *second* round of processing (the remaining) events in S. Specifically, it purges the only event $\{(\langle T_6, T_7\rangle, c, P_{11}, t_{22})\}$ left in S, inserts it to the front of S_T, and then issues another recursive call *Trace*(T_6), which generates the bottom portion of the table. The status of updated S, updated S_T, and recursive calls in the call *Trace*(T_7) which are associated with the two rounds of processing events in S, are marked by the symbol | in Table 5.3.

The final result of S_T, returned by the call *Trace*(T_7), is: $\{(\langle T_6, T_6\rangle, o, P_8, t_{24}), (\langle T_6, T_6\rangle, i, P_9, t_{25}), (\langle T_6, T_6\rangle, o, P_9, t_{26}), (\langle T_6, T_7\rangle, c, P_{11}, t_{22}), (\langle T_1, T_1\rangle, o, P_1, t_2), (\langle T_1, T_1\rangle, i, P_3, t_5), (\langle T_1, T_1\rangle, o, P_3, t_6), (\langle T_1, T_1\rangle, i, P_4, t_7), (\langle T_1, T_3\rangle, s, P_4, t_8), (\langle T_3, T_3\rangle, o, P_4, t_{11}), (\langle T_3, T_3\rangle, i, P_5, t_{12}), (\langle T_3, T_3\rangle, o, P_5, t_{13}), (\langle T_3, T_3\rangle, i, P_{10}, t_{18}), (\langle T_3, T_3\rangle, o, P_{10}, t_{19}), (\langle T_3, T_3\rangle, i, P_{11}, t_{21}), (\langle T_3, T_7\rangle, c, P_{11}, t_{22}), (\langle T_7, T_7\rangle, o, P_{11}, t_{23})\}$.

5.6 Conclusion

Finding RFID tags' behavioral discrepancies by examining RFID tag event records is an important and effective method for detecting cloned tags in RFID-enabled supply chains. However, most event models for RFID tags deal with simple events only and thus fail to characterize the tag events in real-world RFID supply chains. We have in this chapter provided a new event model and the corresponding event-tracing algorithm for RFID supply chains. While the newly proposed event model is able to successfully capture some of the complicated event scenarios in real-world RFID supply chains, the event-tracing algorithm designed for the newly proposed event model provides a foundation for future more powerful clone detection techniques that aim to tackle the cloned tag detection problem more effectively.

An immediate future research work would be to extend the existing event-record-based clone detection techniques by integrating the work of this chapter into them.

References

[1] D. Zanetti, L. Fellmann, and S. Capkun, "Privacy-preserving clone detection for RFID-enabled supply chains," in *2010 IEEE International Conference on RFID*, April 2010, pp. 37–44.

[2] D. Zanetti, S. Capkun, and A. Juels, "Tailing rfid tags for clone detection," in *Network and Distributed System Security Symposium (NDSS)*, April 2013.

[3] J. Huang, X. Li, C. Xing, W. Wang, K. Hua, and S. Guo, "Dtd: A novel double-track approach to clone detection for RFID-enabled supply chains," *IEEE Transactions on Emerging Topics in Computing*, in press.

[4] J. Khor, W. Ismail, and M. Rahman, "Prevention and detection methods for enhancing security in an RFID system," *International Journal of Distributed Sensor Networks*, vol. 2012, Article ID 891584, 8 pages, 2012. doi:10.1155/2012/891584.

[5] W. Xin, "Research on the security and privacy issues in RFID-based supply chains," Ph.D. dissertation, Peking University, 2013.

[6] V. Hinkka and J. Tatila, "RFID tracking implementation model for the technical trade and construction supply chains," *Automation in Construction*, vol. 35, no. 1, pp. 405–414, 2013.

[7] F. Kerschbaum and N. Oertel, "Privacy-preserving pattern matching for anomaly detection in RFID anti-counterfeiting," in *Proceedings of the 6th International Workshop on RFID Security*, 2010, pp. 124–137.

[8] M. Lehtonen, F. Michahelles, and E. Fleisch, "How to detect cloned tags in a reliable way from incomplete RFID traces," in *Proceedings of the 2009 IEEE International Conference on RFID*, 2009, pp. 257–264.

[9] T. Mackey and B. Liang, "The global counterfeit drug trade: patient safety and public health risks," *Journal of Pharmaceutical Sciences*, vol. 100, no. 11, pp. 4571–4579, 2011.

[10] K. Koscher, A. Juels, V. Brajkovic, et al., "Epc rfid tag security weaknesses and defenses: Passport cards, enhanced drivers licenses, and beyond," in *Proceedings of the 2009 ACM Conference on Computer and Communications Security*, 2009, pp. 33–42.

[11] M. Lehtonen, D. Ostojic, A. Ilic, and F. Michahelles, "Securing RFID systems by detecting tag cloning," in *Pervasive Computing*, Lecture Notes in Computer Science, H. Tokuda *et al.*, editors, 2009, vol. 5538, pp. 291–308.

[12] J. Abawajy, "Enhancing RFID tag resistance against cloning attack," in *Third International Conference on Network and System Security*, 2009, pp. 18–23.

[13] Q. L. Xiang, P. Q. Zhang, D. S. Ouyang, and C. H. Feng, "Multiple frequency slots based physical unclonable functions," *Journal of Electronics and Information Technology*, vol. 34, no. 8, 2012, pp. 2007–2012.

[14] H. S. Kou, Z. N. Zhang, and J. Ma, "Physical unclonable function based RFID mutual authentications," *Computer Engineering*, vol. 39, no. 6, 2013.

[15] T. Dimitriou, "A lightweight rfid protocol to protect against traceability and cloning attacks," in *First International Conference on Security and Privacy for Emerging Areas in Communications Networks*, 2005, pp. 59–66.

[16] Z. H. Niu and X. H. Wu, "Physical unclonable function based RFID mutual authentications," *Chinese Journal of Engineering Mathematics*, vol. 27, no. 5, 2010.

[17] J. Shen, D. Choi, S. Moh, and I. Chung, "A novel anonymous RFID authentication protocol providing strong privacy and security," in *2010 International Conference on Multimedia Information Networking and Security*, 2010, pp. 584–588.

[18] L. Mirowski and J. Hartnett, "Deckard: A system to detect change of RFID tag ownership," *Journal of Computer Science and Network Security*, vol. 7, no. 7, pp. 89–98, 2007.

[19] K. Bu, X. Liu, J. Luo, *et al.*, "Unreconciled collisions uncover cloning attacks in anonymous rfid systems," *IEEE Transactions on Information Forensics and Security*, vol. 8, no. 3, pp. 429–439, March 2013.

[20] J. Shi, S. M. Kywe, and Y. J. Li, "Batch clone detection in RFID-enabled supply chain," in *2014 IEEE International Conference on RFID*, 2014, pp. 118–125.

[21] B. J. Albert and C. Li, "The radio frequency identification tag with the function of anti-collision and anti-counterfeiting," in *Proceedings of the Ninth International Conference on Machine Learning and Cybernetics*, 2010, pp. 2681–2686.
[22] S. M. Kywe, Y. Li, and J. Shi, "Attack and defense mechanisms of malicious EPC event injection in EPC discovery service," in *Proceedings of the 2013 IEEE International Conference on RFID-Technologies and Applications*, 2013, pp. 1–6.
[23] W. Zhao, X. Y. Liu, S. K. Zhang, and L. F. Wang, "A double random number based secure communication mechanism of RFID discovery service," *Chinese Journal of Electronics*, vol. 41, no. 1, pp. 153–160, 2013.
[24] J. Shi, D. Sim, Y. Li *et al.*, "A secure EPC discovery services system in EPCglobal network," in *Proceedings of the second ACM conference on Data and Application Security and Privacy*, 2012, pp. 267–274.
[25] J. Worapot, Y. Li, and A. Somjit, "Design and implementation of the EPC discovery services with confidentiality for multiple data owners," in *2010 IEEE International Conference on RFID-Technology and Applications*, 2010, pp. 19–25.
[26] D. Kim, J. Kim, and S. Lee, "Design and implementation for EPC system method to authentication and cryptography," in *Proceedings of the 2008 International Conference on Information Security and Assurance*, 2008, pp. 137–141.
[27] F. Kerschbaum, "An access control model for mobile physical objects," in *Proceedings of the 15th ACM symposium on Access control models and technologies*, 2010, pp. 193–202.

Chapter 6
A new clone detection approach in RFID-enabled supply chains

Cong-cong Xing[1], *Jun Huang*[2], *Kun Hua*[3], *and Song Guo*[4]

Abstract

We propose a new method for detecting cloned tags in RFID-enabled supply chains, aiming to overcome the deficiencies of the existing clone detection techniques in RFID supply chains, caused in partial by the dynamic changes and incidents in radio frequency identification (RFID) supply chains. We first provide a formal (categorical) model for the activities in RFID supply chains, and then present a new cloned tag detection technique by adding a verification bit into tag events, which is followed by some clone detection examples. Next, we compare the performance of the newly proposed clone detection technique against Zanetti's clone detection mechanism by conducting relevant experiments, and finally discuss the related work in the literature. The results of the experiments demonstrate that our proposed mechanism is effective, reasonable, and outperforms Zanetti's work in terms of hit rate.

6.1 Introduction

In radio frequency identification (RFID)-enabled supply chains, a unique product identifier, namely, the electronic product code (EPC), is tagged on each product. Since EPC-related information of products is stored in the local EPC Information Services (EPCIS) located at each supply chain participant and is processed individually, various supply chain participants may record, store, and share their product-related information by leveraging RFID infrastructures (e.g., EPC-global Network).

[1]Department of Mathematics/Computer Science, Nicholls State University
[2]Institute of Electronic Information and Networking, Chongqing University of Posts and Telecommunications
[3]Department of Electrical and Computer Engineering, Lawrence Technological University
[4]School of Computer Science and Engineering, The University of Aizu

Although the implementation of a transparent and real-time supply chain management system and the improvement of warehouse management efficiency for logistics enterprises can be achieved by employing the RFID technology, it, unfortunately, also brings in several issues. For example, cloned tags, once in the hands of criminals and/or terrorists, may endanger the safety of patients in health-care industry [1], threaten the national security and the military force of a country [2], and strike serious economic damages in logistics industries [3–6] leading to consumers' interest and property losses. As such, various clone attack prevention and clone attack detection techniques have been studied aiming to effectively address these issues.

Clone attack prevention: Clone attack prevention methods are cryptography-based techniques (such as encryption, decryption, and authentication) and are composed of (cryptographic) key distributions and management policies. Unfortunately, these prevention techniques, due to their requirements for extra storage spaces and additional encryption operations [3,7], are not suitable for being deployed in those low-cost RFID tags that typically have very low computational power.

Clone attack detection: As no clone attack prevention strategies can completely prevent cloned tags from transpiring, clone attack detection techniques will play a significant and complement role. When the security system of a supply chain is shut down or weakened, counterfeiters can inject a large number of cloned tags into the supply chain. In this case, clone attack detections are the only way to protect the interests of consumers, and thereby are the foundation of the security infrastructure. Clearly, the importance of studying clone detection techniques cannot be overstated.

This chapter proposes an efficient clone detection approach, coined double-track clone detection (DTD). The proposed approach stores a verification value ν in a tag's memory, which is subsequently updated to $\nu + 1$ after the tag is read by the reader. This verification value is part of the data associated with any event that occurs at any supply chain participant. For the sake of privacy protection, the initial ν-value is set to be a random number (within permitted range). We assume that the ν-value is 8-bit long, the EPC of a tag cannot be modified, but the memory of a tag can be read and rewritten. All ν-values of a tag form a verification sequence as the tag transits along the supply chain. While the ν-value sequence constitutes one track of product information, business transaction information of events forms another. The newly proposed clone detection scheme checks the correctness of both tracks with reference to specific tag events. Note that even if the ν-value is modified by attackers, the proposed approach can still detect clones because the modification will probably cause duplicated ν-values. Due to its independence of a predefined structure of supply chains and on the information flow of products, the proposed DTD approach not only fits well to supply chains that change dynamically, but is suitable for general deployment as well.

The remainder of this chapter is organized as follows. Section 6.2 presents a formal (categorical) characterization of product flows in RFID-enabled supply chains. Our proposed clone detection approach is introduced in Section 6.3, and is evaluated in Section 6.4 with comparison to Zanetti's work. Section 6.5 discusses the related work in the literature and Section 6.6 concludes this chapter.

6.2 A categorical perspective of RFID supply chains

In RFID-enabled supply chains, each product is labeled with an RFID tag which contains an EPC, and an RFID tag and the product to which it is attached are considered inseparable. So throughout the chapter, the terms RFID tags and EPC tags and the terms cloned tags and cloned products are both used interchangeably.

The EPC contained in a product is to be read (multiple times) by RFID readers at different supply chain participants' locations. Each tag-reading at a location creates an event (see formal definition below) which is stored in the local EPCIS database. The created event is a log of a tag reading activity, consists of various pieces of information related to the tag reading, and can be accessed and shared by all (allowed) RFID supply chain partners. As such, all events associated with a specific tag during its lifetime in the supply chain are stored in a distributed manner. Discovery Services (DS) is a central repository database which is connected to all local EPCIS databases, and may, upon requests, create a product event history path for each product by accessing various distributed EPCIS databases. Supply chain participants, such as manufacturers, wholesalers, as well as retailers, may access the data in DS through authentication and access control mechanisms. We assume that all supply chain participants are legitimate and act in good faith.

Below, we give a formal definition of tag events. An event, $e(id, t)$, which is created for reading a product (identified by its $id =$ EPC) at time t in an RFID supply chain, is defined as follows:

$$
\begin{aligned}
Id &= \text{set of tag identifications (EPCs)} \\
Tm &= \text{set of times} \\
L &= \text{set of locations of supply chain participants} \\
Tr &= \{r, s, i\} \\
V &= \{0, \ldots, 255\} \\
S &= \{tr, fs\} \\
Ev &= Id \times Tr \times L \times Tm \times V \times S \\
e &: Id \times Tm \rightarrow Ev \\
e(id, t) &= (id, \tau, \ell, t, \nu, \sigma) \in Id \times Tr \times L \times Tm \times V \times S
\end{aligned}
$$

where the meanings of the sets Id, Tm, and L are as explained, and Tr, V, S, and Ev denote the set of business transactions (receiving (r), shipping (s), and inventory (i)), the set of verification values, the set of status (i.e., the success (tr) or failure (fs) of updating the verification value in an event), and the set of events, respectively. e is a function with domain $Id \times Tm$ and codomain Ev, taking a pair of product identification and time and returns the corresponding event.

Two special events $e(id, t_{in})$ and $e(id, t_{out})$ are created for a product when the product initially enters into the supply chain (i.e., when an EPC tag is assigned to a product at the manufacturer) and eventually leaves the supply chain (i.e., the product is sold at a retailer). So cloned products can be easily detected using the corresponding events if they appear on the supply chain before $e(id, t_{in})$ or after $e(id, t_{out})$. An event is considered to be proprietary and confidential. Any supply chain participants only know their direct business partners, and can join or leave the supply chain at any time. We define clones as counterfeit products carrying legitimate EPCs, and multiple reading results of one tag are assumed to be processed during the data collection stage.

The activities associated with the clone detection in RFID-enabled supply chains can be understood and formalized as an Olog model [8] as partially[1] shown in Figure 6.1, in which each box represents an entity type and each arrow denotes a mathematical function from the source box (domain) to the target box (codomain). Note that there are multiple arrows going from box E into box D, but only three arrows are drawn in the diagram for the sake of succinctness. The check mark ($\surd$) indicates that the enclosing figure is *commutative*, i.e., any two paths leaving from the same source box and ending at the same target box are equivalent in the sense that they yield the same result. For example, the check mark in the lower right triangle ABC explicitly states that the EPC which is being read in a tag-reading is exactly the same EPC of the product for which the tag-reading action is performed.

In fact, the Olog diagram shown in Figure 6.1 precisely constitutes a mathematical category [8,9]. Let us call this category **Tag**, and give a brief explanation (or proof) as follows.

- *Ob*(**Tag**): Each box in Figure 6.1 represents an object in category **Tag**. They are actually sets of the corresponding entities. For example, box A in Fig 6.1 is the set of *all* products.
- *Hom*(**Tag**): Morphisms in category **Tag** are functions (represented by arrows) in Figure 6.1. Specifically, for any pair of objects a and b in **Tag**, $Hom_{\mathbf{Tag}}(a, b)$ is the set of functions depicted in Figure 6.1 and their compositions with a being domain and b being codomain. For example, the edge from box A to box B labeled by "has" in Figure 6.1 is a function that sends a product to its unique *EPC*.
- Identify morphisms: For each object $a \in \mathbf{Tag}$, the identify morphism id_a for a is just the identity function from a to itself. In terms of Olog (where functions, instead of being abstract, are always labeled with English phrases or sentences), the identity morphism of an object can be understood as or labeled by "doing nothing." Note that these identity functions are not shown in Figure 6.1 due to their trivialness.
- Composition operation: For any morphisms

  $$f \in Hom_{\mathbf{Tag}}(a, b) \quad \text{and} \quad g \in Hom_{\mathbf{Tag}}(b, c),$$

  we have $g \circ f \in Hom_{\mathbf{Tag}}(a, c)$, where $\circ$ is just the (regular) function-composition operation. For example, the commutativity (marked by the check sign $\surd$) of the lower right triangle ABC in Figure 6.1 suggests that the function labeled by "directly read" is the composition of the two functions labeled by "is performed

[1]Some Olog edges are not drawn in the diagram in order not to clutter it.

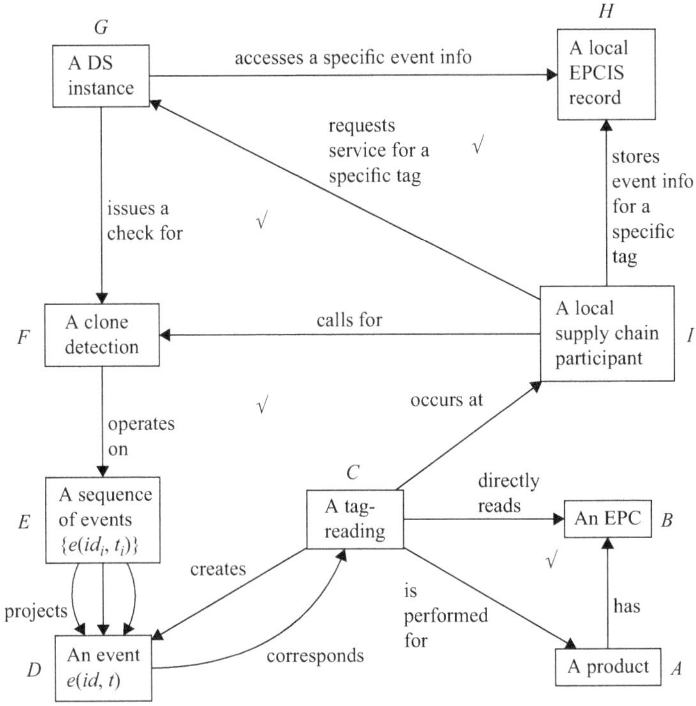

Figure 6.1 The Olog model for RFID-enabled supply chain activities

for" and "has." In other words, if we rename these three functions as "*dr*," "*ipf*," and "*h*," respectively, then we have

$$dr = h \circ ipf.$$

- Identity law: For any identity function id_a of object $a \in Ob(\mathbf{Tag})$, and any morphisms $f \in Hom_{\mathbf{Tag}}(a, b)$ and $g \in Hom_{\mathbf{Tag}}(c, a)$, we have

  $$f = f \circ id_a \quad \text{and} \quad g = id_a \circ g$$

  which is guaranteed by the definition of identity function and the definition of function compositions. So the identity law is satisfied. For instance, looking at the object C and the function labeled by "directly reads" in Figure 6.1, the identity law instance $dr = dr \circ id_C$ can be intuitively understood as: for a tag-reading, the sequence of "doing noting" activity followed by the "directly reads" activity is equivalent to the "directly reads" activity alone.
- Composition law: For any morphisms $f \in Hom_{\mathbf{Tag}(a,b)}$, $g \in Hom_{\mathbf{Tag}(b,c)}$ and $h \in Hom_{\mathbf{Tag}(c,d)}$, we have

  $$h \circ (g \circ f) = (h \circ g) \circ f$$

  which is, again, guaranteed by the definition of function compositions. Hence the composition law is also satisfied. For instance, looking at the three functions

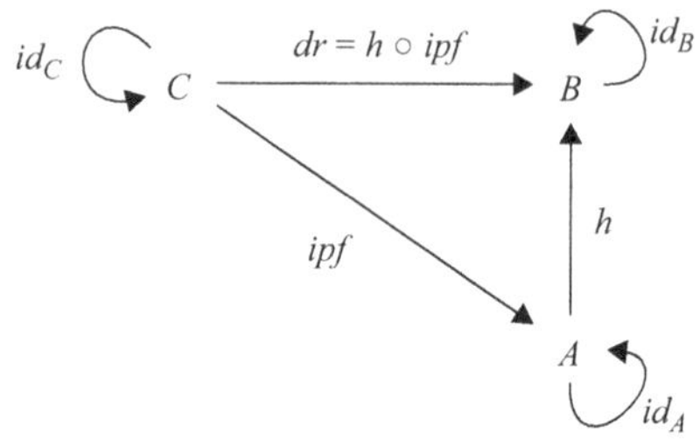

Figure 6.2 Part of the ***Tag*** *category*

labeled by "occurs at," "calls for," and "operates on" in Figure 6.1, the composition law regarding them can be understood as: the sequence of events operated on by a clone detection which is the result of the request of a local supply chain participant where the related tag-reading occurs, is the same sequence of events which is the result of a local supply chain participant's calling for clone detection and the subsequent operation, with the condition that the related tag-reading occurs at this local supply chain participant.

As an illustrative example for the objects and morphisms (including identity morphisms) together with morphism compositions, all objects and morphisms in the category **Tag** that are directly associated with the lower right triangle in Figure 6.1 are depicted in Figure 6.2.

6.3 The clone detection system

A verification sequence of ν-values for the purpose of detecting clones is formed by following some stipulated rules except that the initial item in the verification sequence may be assigned randomly by the manufacturer. Our approach detects the presence of clones by examining, for two consecutive events in time, the successiveness of the ν-values and the consistency of business transactions.

6.3.1 ν-Value verification sequence

A ν-value verification sequence for a product is built up by successively updating the ν-value in the tag of the product. ν-value updating is completed in a non-interactive manner, i.e., by the participating RFID reader alone. It is a natural extension of the tag-reading process in the sense that a new event containing the updated ν-value will be created after the current event (containing an old ν-value) has been read. The procedure of the ν-value updating includes the following steps: (1) Read the EPC and the ν-value from the tag memory. (2) Increase ν by 1 and write the result back into the tag memory. Tag-writing mistakes are indicated by the status attribute σ. When the reader does not receive an acknowledging response from the tag for the writing operation, or the writing operation fails, σ will be set to f. (3) Create an event $e(id, t) = (id, \ell, \tau, t, \nu, \sigma)$, and add it to the local database. Of course, supply chain participants must agree to the above specifications. In addition, the reader is capable of signaling a request at any time to disable the σ-attribute.

6.3.2 Event track formation

Any supply chain participants may request any product-related information from the DS, and the DS can access the distributed EPCIS databases to create a history path of events – an event track – for any product upon the request from the supply chain participants. When requested by supply chain partners, our clone detection approach, as a third-party service program, may be authorized to access, collect, and analyze all event data associated with a particular tag EPC to build the track of events for this tag EPC.

6.3.3 Clone detection rules

All available events associated to a specific tag EPC are collected and ordered by time to form an event path. The machinery of our clone detection approach can be precisely expressed by the following formulated rules:

(1) $e(id,t) = (id, \ell, \tau, t, \nu, \sigma)$
(2) For any time t_i, if $e(id,t_i)_\tau = r$,
then $e(id,t_{i+1})_\tau = s \mid i$ and $e(id,t_i)_\ell = e(id,t_{i+1})_\ell$;
(3) For any time t_i, if $e(id,t_i)_\tau = s$,
then $e(id,t_{i+1})_\tau = r$ and $e(id,t_i)_\ell \neq e(id,t_{i+1})_\ell$;
(4) For any time t_i, if $e(id,t_i)_\tau = i$,
then $e(id,t_{i+1})_\tau = i \mid s$ and $e(id,t_i)_\ell = e(id,t_{i+1})_\ell$;
(5) For any time t_i, if $e(id,t_i)_\sigma = tr = e(id,t_{i+1})_\sigma$,
then $e(id,t_{i+1})_\nu - e(id,t_i)_\nu \equiv 1 \pmod{256}$;

where t_i and t_{i+1} represent two arbitrary consecutive points in time, $e(id,t)_\ell$ (respectively, $e(id,t)_\tau$, $e(id,t)_\nu$, $e(id,t)_\sigma$) signifies the ℓ (respectively, τ, ν, σ) component of the $e(id,t)$ tuple, and $s \mid i$ means s or i.

Formula (1) is just a redisplay of the event formulated in Section 6.3. Following the standard format of inference rules in logic, all rules are written in the form of "if A then B" clause where A represents the premise and B represents the conclusion. Specifically, rule (2) states that for any given event, if the business transaction of this event is "receiving" ($e(id,t_i)_\tau = r$), then this event must be followed by a shipping or inventory event recorded at the same location. In a similar fashion, rule (3) stipulates that a shipping event recorded at a location must be followed by a receiving event recorded at a different location, and rule (4) states that an inventory event must be followed by either an inventory event or a shipping event at the same location. Rule (5) states that if the verification values of two time-consecutive events are well documented, then the verification value of the later event is one more than that of the early event modulo 256. We can regroup rules (2)–(5) into two composite rules as follows:

$$\begin{aligned} Rule\ I &= \{rule(2), rule(3), rule(4)\} \quad \text{(business transaction track)}, \\ Rule\ II &= \{rule(5)\} \quad \text{(verification sequence track)}, \end{aligned}$$

and any pair of time-consecutive events passes the clone-check if and only if both Rule I and Rule II are satisfied.

6.3.4 Clone detection examples

We can examine the correctness of all such pairs for any given set of events through the above two (composite) rules, thereby forming a double-track inspection for clones. If all examinations yield correct (pass) results, then there is no presence of clones; otherwise, some cloned products exist in the supply chain. These two situations are illustrated in Figure 6.3(a) and (b), respectively. Figure 6.3(a) shows the detection result for a case where there are no cloned products, and Figure 6.3(b) shows the detection result for a case where there are some cloned products. Note that the successful detection of the clone existence in Figure 6.3(b) would not be possible without Rule II, since Rule I gives a pass to all inspections in this situation. Incidentally, accidents may conceal the presence of clones or create incorrect observations leading to a false alarm. There are three types of accidents: misevent, misread, and miswrite, and their respective effects are illustrated in Figure 6.3(c). For instance, due to the misevent at time t_2 or the miswrite at time t_3, the first examination yields a "fail" and therefore causes a false alarm since there are no clones involved in that examination. The result of the third examination should have been a "fail" since there is a cloned product involved in the examination; but because of the misreading of the clone tag at time t_5, no event is created for this cloned product for that time slot and thus the existence of the cloned product is concealed.

We now address the issue of determining the cause of failure when the double-track rule verification yields a negative result. That is, does the failure suggest the presence of some clone products or is it caused by the combination of misevent, misread, and miswrite? There are basically two approaches to resolving this issue.

Binomial approach: Assume P_{mr} is the misreading probability of the reader. The readings of the tags can be considered binomially distributed as shown in (6.1), where N_m is the total number of missing events that would be required to restore all incorrect sequences in the considered track, and N is the total number of events. If the probability calculated by (6.1) exceeds a certain threshold δ, then the cause of the rule verification failure can be regarded as clones; otherwise, it is due to the combination of misevent, misread, and miswrite.

$$P_{tr} = 1 - \sum_{k=N_m}^{N} C_k^N P_{mr}^k (1 - P_{mr})^{N-k} \tag{6.1}$$

$$P_{tr} > \delta \quad (Clone) \tag{6.2}$$

$$P_{tr} \leq \delta \quad (Read/Write\ error) \tag{6.3}$$

Ratio approach: Let RV_f be the number of failed rule verifications, and RV be the total number of rule verifications. If the ratio of RV_f to RV is over a predefined

Tag id = T_1					
Time	t_1	t_2	t_3	t_4	t_5
Genuine tag event	$(T_1, r, l_1, t_1, 10, t)$	$(T_1, i, l_1, t_2, 11, t)$	$(T_1, s, l_1, t_3, 12, t)$	$(T_1, r, l_2, t_4, 13, t)$	$(T_1, s, l_2, t_5, 14, t)$
Location	l_1	l_1	l_1	l_2	l_2
B. track	r	i	s	r	s
V. track	10	11	12	13	14
Rule I	Pass	Pass	Pass	Pass	
Rule II	Pass	Pass	Pass	Pass	
Result	Pass	Pass	Pass	Pass	

(a)

Tag id = T_2						
Time	t_1	t_2	t_3	t_4	t_5	t_6
Genuine tag event	$(T_2, r, l_1, t_1, 10, t)$	$(T_2, i, l_1, t_2, 11, t)$		$(T_2, s, l_1, t_4, 12, t)$	$(T_2, r, l_2, t_5, 13, t)$	$(T_2, s, l_2, t_6, 14, t)$
Cloned tag event			$(T_2, i, l_1, t_3, 12, t)$			
Location	l_1	l_1	l_1	l_1	l_2	l_2
B. track	r	i	i	s	r	s
V. track	10	11	12	12	13	14
Rule I	Pass	Pass	Pass	Pass	Pass	
Rule II	Pass	Pass	Fail	Pass	Pass	
Result	Pass	Pass	Fail	Pass	Pass	

(b)

Tag id = T_3						
Time	t_1	t_2	t_3	t_4	t_5	t_6
Genuine tag event	$(T_3, r, l_1, t_1, 10, t)$	$(T_3, i, l_1, t_2, 11, t)$	$(T_3, s, l_1, t_3, 10, t)$	$(T_3, r, l_2, t_4, 13, t)$		$(T_3, s, l_2, t_6, 14, t)$
mr/w/e		Misevent	Miswrite			
Cloned tag event					$(T_3, i, l_2, t_5, 13, t)$	
mr/w/e					Misread	
Location	l_1		l_1	l_2		l_2
B. track	r		s	r		s
V. track	10		10	13		14
Rule I		Pass	Pass		Pass	
Rule II		Fail	Fail		Pass	
Result		Fail	Fail		Pass	

(c)

Figure 6.3 Clone detections. (a) Results of a genuine product. (b) Results of a genuine product mixed with a cloned product. (c) Results of a genuine product mixed with a cloned product with misevent, misread, and miswrite being considered (B. track = business transaction track; V. track = verification value track; mr/w/e = misread, miswrite, misevent; t = tr)

threshold ϕ, then the cause of the failure is probably cloned tags; otherwise, the cause can be attributed to reading/writing mistakes. That is,

$$R = \frac{VR_f}{VR} \tag{6.4}$$

$$R > \phi \quad (Clone) \tag{6.5}$$

$$R \leq \phi \quad (Read/Write\ error) \tag{6.6}$$

The probability method and the ratio method can be coupled with double-track sequence verifications to determine the presence of clones. The specific steps are as follows:

(i) Use Rule I and Rule II to determine whether a given pair of time-consecutive events is in the correct order; the answer is positive if and only if both Rule I and Rule II are satisfied. There would be no presence of cloned products in the given set of events if all examinations for all time-consecutive pairs of events yield a positive answer.
(ii) When the result obtained from step (i) is negative, use (6.1)–(6.3) or (6.4)–(6.6) to determine the possibility of the presence of clones.

6.4 Evaluation and comparison with peer work

In this section, we evaluate our clone detection scheme by conducting simulation experiments. A 15-partner supply chain in the form of a FOUR-level binary tree is constructed by using the Arena [10] simulation software (see Figure 6.4). Products flow in the supply chain from the manufacturer to retailers via one or several distributors. Our clone detection approach will be triggered when a genuine or counterfeit product leaves the supply chain (is sold to customers). The manufacturer (top level) produces 1 000 genuine products every day, and 10 cloned products are randomly injected into different levels in the supply chain on a daily basis.[2] Specific parameter settings of the simulation are shown in Table 6.1.

The following aspects regarding our clone detection approach are evaluated: storage space requirement, computation workload, communication cost, and the clone detection rate. The use of EPC C1G2 RFID tags is stipulated in the simulation experiments.

Storage space requirement: Only a small amount (8 bits) of storage space is required for the ν-value which can be generally met by any (and even low-cost) tags. Our scheme will not increase the number of events in the local database. Compared with [12], the set of relevant attributes of each event is extended by ν and σ, resulting in only 7% increases in event size.

[2]The clone product injection rate is stipulated to be 1% in the simulation. By the way in which the DTD works, if the clone injection rate is higher than 1%, then the clone detection rate would not be less than the current clone detection rate.

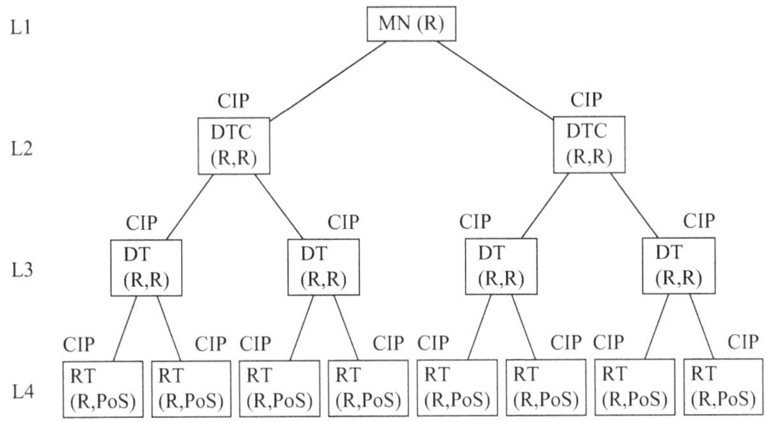

Figure 6.4 A four-level 15-partner binary tree supply chain (MN = manufacturer, DTC = distribution center, DT = distributor, RT = retailer, R = reader, PoS = point of sale, CIP = clone injection point, Li = level i). (Adapted from Zanetti et al. [11])

Table 6.1 Simulation parameters

Parameter	**Value**
Misread probability(P_{mr})	$N(5\%, 1\%)$
Miswrite probability(P_{mw})	$N(5\%, 1\%)$
Misevent probability(P_{me})	$N(5\%, 1\%)$
Production rate for genuine products	1 000 products/day
Production rate for counterfeit products	10 products/day
Production time	2 months
Shipping time	8 AM every day
Stocking time	$N(3, 0.5)$ days
Transportation time	$N(1, 0.25)$ days
Output load (demand)	Uniformly distributed
Supply-chain structure	FOUR-level binary tree
ν value	8 bits
Counterfeit injection point	Random at any partner

Computation workload: Tags themselves do not perform any computations. Readers only perform a primitive operation (increment by 1 for the ν-value stored in tags), and the rule verifications are simple and lightweight logical operations.

Communication cost: While our clone detection scheme requires the reader perform some extra writing operations, it does not inflict any communication overheads with the local databases at the back end. Also, compared with the tailing mechanism in Reference 11 which works on 3 bytes of data (16 bits for the tail and the pointer

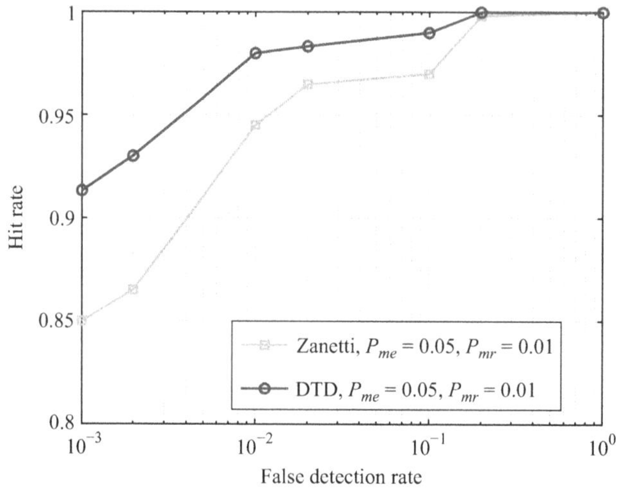

Figure 6.5 Hit rate comparison between Zanetti's work and DTD

and 8 bits for the flag), our scheme induces a simpler communication process that only needs to update the 8-bit ν-value.

Clone detection rate: Note that the presence of clones will not be detected by the work in Reference 12 when the business transactions in cloned products and in genuine products are consistent. This issue is well resolved in our clone detection scheme by enforcing the consecutiveness of the ν-values in two adjacent events in addition to the requirement of business transaction consistency. Clearly, in theory, clones can still be potentially detected by our DTD scheme even if the business transactions in cloned products and in genuine products do not show any evidence of counterfeits, which will lead to a higher rate of clone detections. This theoretical observation is verified by the experimental results: the hit rate of our clone detection scheme is 91.3% when the misread probability $P_{mr} = 0.01$, misevent probability $P_{me} = 0.05$, and the false detection rate $FDR = 0.001$, as shown in Figure 6.6. This is a 6% increase over Zanetti's work [12] where the same P_{mr}, P_{me}, and FDR are used (see Figure 6.5); moreover, the hit rate of our work is as high as 98.3% when $FDR = 0.04$. Also, Figures 6.6 and 6.7 indicate that misread has a greater impact than misevent on the clone detection rate in our clone detection approach. Note that injecting clones into a lower level (toward the top of the tree) of the chain generates fewer events than injecting clones into an upper level (toward the bottom of the tree), and for the same number of failed incidents on the double-track checking of adjacent event pairs, a shorter event trace gives a clearer evidence on the presence of clones than a longer event trace. So, for a fixed false detection rate, the hit rate when clones are injected into a lower level is higher than that when clones are injected into an upper level, which is shown in Figure 6.8.

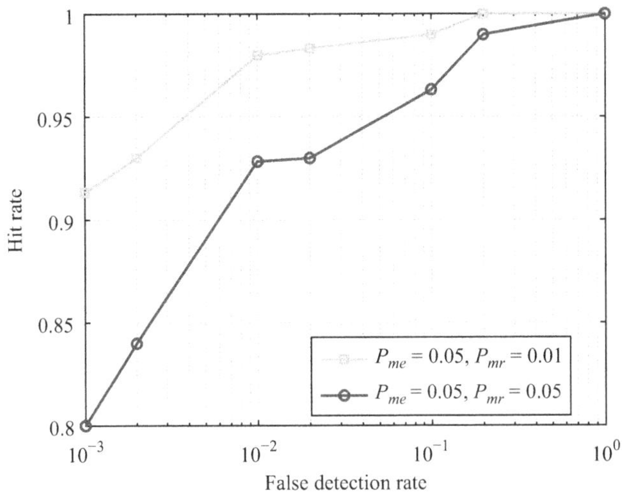

Figure 6.6 Correlations between P_{mr} and hit rate with respect to FDR

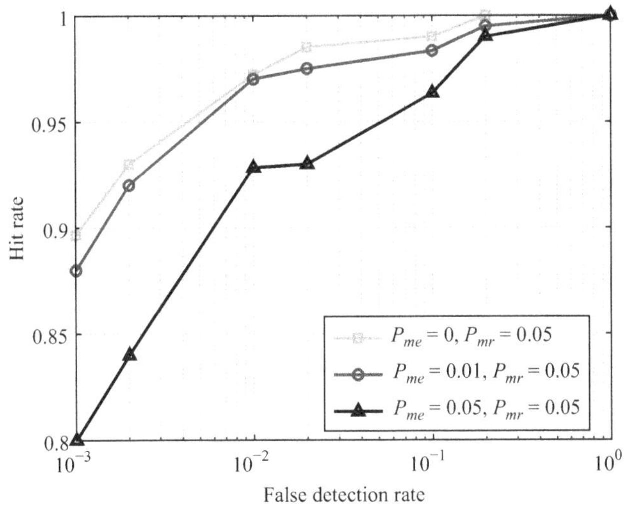

Figure 6.7 Correlations between P_{me} and hit rate with respect to FDR

For convenience, comparisons between our DTD strategy and Zanetti's work [12] are summarized in Table 6.2. In short, our proposed DTD approach outperforms Zanetti's work [12] in term of clone detection rate under the same testing environment and with the same parameter settings, at the cost of slight increase in tag memory and decrease in tag processing speed.

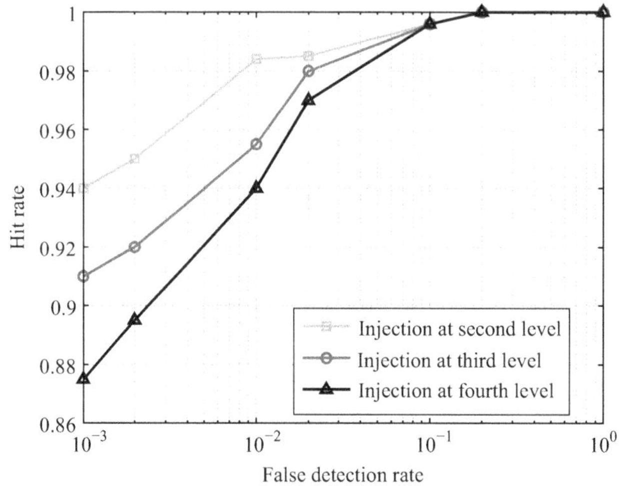

Figure 6.8 Correlations between clone injection level and hit rate with respect to FDR

Table 6.2 Comparison between DTD and Zanetti's work

Aspect	**Zanetti's work**	**DTD**
Tag memory	96 bits for EPC	EPC (96)+ ν-value (8)
Extra EPCIS storage	None	7%
Communication cost	Low	Relative low
Tag processing speed	No effect on speed	Slow down speed
Clone detection rate	Low	High
Pragmatics	General deployment	General deployment

6.5 Related work

Extensive investigations on the cloned products issue in RFID supply chains have been done in the literature. We in this section review some of these studies.

Staake *et al.* [13] presented a preliminary study on the supply chain RFID security solutions in terms of track-and-trace, highlighting the negative impact of incomplete tracks on cloning attack detections when partners do not record or share the track data. Mirowski and Hartnett [14] used statistical anomaly to detect clones by checking the status of the RFID tag ownership, operations of readers, tags and reader IDs, and the time stamp marks of events. Tag paths (the sequence of visited readers) are verified by the data saved in tag memory in References 15 and 16. Lee and Bang [17] proposed a pattern mining algorithm, using event track records to mine the legitimate supply chain model by which a counterfeit product detection algorithm can be generated. Although these proposed mechanisms are all suitable for the low-cost EPC C1G2 tags [18,19], they need the related supply chain structure and product flow information in order to work properly, and thus will result in some weak performance and less robustness when the situation is complicated with supply chain dynamic changes, product recalls, and product transportation errors.

Zanetti *et al.* [12] proposed a track-and-trace-based privacy-preserving clone detection method, which detects clones by verifying the correctness of two consecutive events in time, without relying on the global knowledge of supply chain structures or the product flow information. This method works well with product recalls and product delivery errors; however, the clone detection rate is not improved with this method. A pattern-matching approach was proposed in Reference 20 by Kerschbaum and Oertel to detect illegal transactions between supply chain partners. In Reference 11, Zanetti *et al.* proposed to add a random tail and a tail pointer in each user-defined block in EPC tags. In each event, the reader increments the tail pointer and updates the pointed random bits. Cloned products can be detected by inspecting the consistency between tails and tail pointers. Although enjoying a relatively high detection rate, this method reduces the tag processing speed, and induces a large amount of communication and memory overheads. Bu *et al.* [5,7] suggested the use of hash functions in detecting clones. Under their schemes, two tags with the same ID always simultaneously responds to the reader queries when they are within the reading range of the reader, resulting in the situation that genuine tags and cloned tags will make inevitable irreconcilable collisions. Because of its requirement that genuine tags and clone tags must be present at the same time and in the location, this method can be used only in certain restricted scenarios.

The DTD approach proposed in this chapter simply adds a new and effective clone detection mechanism to the existing ones in the literature, making a step forward toward finding the optimal clone detection technique for RFID supply chains.

6.6 Final remarks

Cloned tags/products are a serious issue in RFID supply chains, and conventional clone detection techniques are inadequate in the sense that they rely on the global structure of supply chains but the supply chains may change dynamically. Toward addressing the deficiencies of the conventional cloned products detection techniques, we have proposed a simple, effective, event-record-based new clone detection technique that considers both the consistency of business transactions and the consecutive nature of the proposed ν-value. We argue that this chapter makes the following contributions.

- The newly proposed technique does not depend on the structure of supply chains and thus makes it universally applicable.
- The simplicity of the proposed technique makes it practically feasible.
- The newly proposed technique has a competitively high clone detection rate with a comparably low communication overhead.

- The newly proposed clone detection technique successively captures the cloned products that will be inevitably overlooked in Zanetti's work [12]. The verification values of events in the proposed method form a natural number sequence, and the strictness and successiveness of natural numbers (i.e., a natural number n has a unique successor $n + 1$ and a unique predecessor $n - 1$) essentially eliminates any chance for any counterfeit products to exist and remain undetected in an RFID supply chain.

Possible future work along this line of research would include investigations on the clone detection techniques where RFID readers may be hijacked and/or supply chain participants may not act in good faith.

References

[1] E. Lefebvre, L. Castro, and L. A. Lefebvre, "Prevailing issues related to RFID implementation in the healthcare sector," in *Proceedings of the 10th WSEAS International Conference on Applied Computer and Applied Computational Science (ACACOS)*, Venice, Italy, March 2011, pp. 266–272.

[2] K. Koscher, A. Juels, V. Brajkovic, and T. Kohno, "EPC RFID tag security weaknesses and defenses: Passport cards, enhanced drivers licenses, and beyond," in *Proceedings of the 16th ACM Conference on Computer and Communications Security*, Nov 2009, pp. 33–42.

[3] J. H. Khor, W. Ismail, and M. G. Rahman, "Prevention and detection methods for enhancing security in an RFID system," *International Journal of Distributed Sensor Networks (IJDSN)*, vol. 2012, Article ID 891584, 8 pages, 2012. doi:10.1155/2012/891584.

[4] Z. D. Sun and J. D. Sun, "RWMS: RFID based weapon management system," *Advanced Materials Research*, Vols. 452–453, pp. 386–390, 2012.

[5] K. Bu, X. Liu, J. Luo, B. Xiao, and G. Wei, "Unreconciled collisions uncover cloning attacks in anonymous RFID systems," *IEEE Transactions on Information Forensics and Security*, vol. 8, no. 3, pp. 429–439, 2013.

[6] M. Lehtonen, F. Michahelles, and E. Fleisch, "How to detect cloned tags in a reliable way from incomplete RFID traces," in *2009 IEEE International Conference on RFID*, Orlando, FL, USA, April 2009, pp. 257–264.

[7] K. Bu, X. Liu, and B. Xiao, "Fast cloned-tag identification protocols for large-scale RFID systems," in *2012 IEEE 20th International Workshop on Quality of Service (IWQoS)*, Coimbra, Portugal, June 2012, pp. 1–4.

[8] D. Spivak, *Categroy Theory for the Sciences*. MIT Press, Cambridge, MA, USA, 2014.

[9] S. M. Lane, *Categories for the Working Mathematician*. Springer, New York, NY, USA, 1998.

[10] A. Memari, A. Anjomshoae, M. Galankashi, and A. Bin Abdul Rahim, "Scenario-based simulation in production-distribution network under demand uncertainty using arena," in *2012 7th International Conference on Computing*

and Convergence Technology (ICCCT), Seoul, Korea (South), December 2012, pp. 1443–1448.

[11] D. Zanetti, S. Capkun, and A. Juels, "Tailing RFID tags for clone detection," in *Network and Distributed System Security Symposium (NDSS)*, San Diego, CA, USA, April 2013.

[12] D. Zanetti, L. Fellmann, and S. Capkun, "Privacy-preserving clone detection for RFID-enabled supply chains," in *2010 IEEE International Conference on RFID*, April 2010, Orlando, FL, USA, pp. 37–44.

[13] T. Staake, F. Thiesse, and E. Fleisch, "Extending the EPC network – The potential of RFID in anti-counterfeiting," in *ACM Symposium on Applied Computing (SAC 2005)*, March 2005, Santa Fe, NM, USA, pp. 1607–1612.

[14] L. Mirowski and J. Hartnett, "Deckard: A system to detect change of RFID tag ownership," *International Journal of Computer Science and Network Security*, vol. 7, no. 7, pp. 89–98, 2007.

[15] E.-O. Blass, K. Elkhiyaoui, and R. Molva, "Tracker: Security and privacy for RFID-based supply chains," in *18th Annual Network and Distributed System Security Symposium (NDSS 2011)*, San Diego, CA, USA, February 2011.

[16] K. Elkhiyaoui, E.-O. Blass, and R. Molva, "Checker: On-site checking in RFID-based supply chains," in *Proceedings of the Fifth ACM Conference on Security and Privacy in Wireless and Mobile Networks*, Tucson, AZ, USA, April 2012, pp. 173–184.

[17] H. S. Lee and H. C. Bang, "Detecting counterfeit products using supply chain event mining," in *2013 15th International Conference on Advanced Communication Technology (ICACT)*, PyeongChang, Korea (South), January 2013, pp. 744–748.

[18] http://www.impinj.com. Retrieved in April 2016.

[19] http://www.rfidchina.org. Retrieved in April 2015.

[20] F. Kerschbaum and N. Oertel, "Privacy-preserving pattern matching for anomaly detection in RFID anti-counterfeiting," in *Proceedings of the 6th International Conference on Radio Frequency Identification: Security and Privacy Issues*, Istanbul, Turkey, June 2010, pp. 124–137.

Chapter 7
Participatory sensing network: a paradigm to achieve applications of IoT

Fen Hou[1], *Jingyi Sun*[1] *and Shaodan Ma*[1]

Abstract

Internet of Things (IoT) aims to provide ubiquitous connectivity and services for individual persons and machines. The related applications cover various areas in industry, healthcare, city management, etc. The efficient information collection is very critical for the success of these applications. Without enough number of data, it is difficult for a system to provide high-quality services. Participatory sensing network (PSN) is a promising paradigm to efficiently collect information/sensing data. Meanwhile, the incentive mechanism design plays a key role in achieving the collection of enough number of sensing data in PSN, where the sensor-equipped mobile devices are owned and controlled by individual users. Most of existing works on incentive mechanism design focus on the participation of smartphone users, rather than the quality of sensing data. However, data quality is also an important factor for a data collector since smartphone users may submit the erroneous or unreliable data. Low-quality data will impact the accuracy of data analysis result and degrade the provided service. Therefore, data quality should be considered in the incentive mechanism design. Reputation is one way to evaluate the quality of data provided by a mobile user. In this chapter, we introduce the significance of incentive mechanism design for the applications of IoT, then we design a reputation-aware incentive mechanism. Taking the quality of sensing data into account, the proposed mechanism can maximize the weighted social welfare of the whole system and guarantee the nice features of truthfulness and individual rationality. Extensive simulations have been conducted to demonstrate the better performance of the proposed incentive mechanism compared with other existing methods in terms of the weighted social welfare and the average reputation.

7.1 Introduction

As a new phase of information society, Internet of Things (IoT) is expected to provide ubiquitous connectivity and services for individual persons and machines.

[1]Department of Electrical and Computer Engineering, University of Macau, Macao

IoT applications cover various areas [1] such as transportation and traffic control, smart healthcare, smart environment, and smart city and society. For example, a smartphone-assisted chronic illness self-management system can collect the real-time biomedical and environmental data, then the system can provide the intelligent healthcare service based on the collected data [2].

To make these applications be successful, the efficient information collection plays one of the most important roles. For instance, without enough data to reflect the traffic status at different areas in a city, the precise and comprehensive analysis and virtualization of the traffic situation cannot be achieved, which makes the provision of smart traffic control or intelligent transportation system be difficult. Participatory sensing network (PSN) is a promising paradigm to efficiently collect information/sensing data from the surrounding environment such as temperature, noise level, and traffic situation. Therefore, it plays a key role in the success of IoT applications.

With the advance of hardware technology, the size of sensor devices becomes smaller and smaller such that various sensors (e.g., GPS, gyroscope, camera, accelerometer) have been widely embedded in mobile devices. Meanwhile, we have witnessed the explosive growth of the mobile users in recent years [3]. The popularity of sensor-embedded mobile devices in our daily life makes PSN become a promising paradigm to collect a large amount of high-quality sensing data in different areas [4,5]. Compared with a traditional fixed sensor network, PSN exists many novelties in both the network structure and the applications due to its good features in terms of mobility and scalability. However, it also poses many new challenging issues in the aspects of mechanism design, resource allocation, and scheduling. Successfully addressing these challenging issues in PSN will enable the success of IoT.

Incentive mechanism design is one of the most critical challenges to achieve the collection of enough sensing data in PSN. In a traditional sensor network, enough sensing data can be easily collected since the sensor network is usually deployed by some specific service provider (SP), which can control the activity of the deployed sensors. However, in PSN, the sensor-equipped mobile devices are owned and controlled by individual users. They may be reluctant to involve in the sensing activity and contribute their sensing data since the process of sensing consumes the resource of their mobile devices (e.g., power, memory, computing capacity). Therefore, how to motivate the rational mobile users to participate in the sensing activity is one of the most significant challenges. If an SP cannot collect enough sensing data, the service quality (e.g., accuracy of air pollution, traffic congestion) cannot be guaranteed.

A few research efforts have been made to address this issue. Lee and Hoh [6] proposed an incentive mechanism using iterative reverse auction, where users could sell their sensing data to an SP with the claimed bid price. The proposed mechanism introduced the virtual participation credit to avoid the cost explosion, maintain adequate number of participants, minimize the incentive cost, and prevent the smartphone users from dropping out of the sensing procedure. However, the designed inventive mechanism cannot guarantee the property of truthfulness. Duan *et al.* [7] designed incentive mechanisms for data acquisition for the collection of enough data to set up a database, where a reward-based collaboration mechanism was proposed to formulate this problem as a two-stage Stackelberg game. In the first stage, the client announced the total reward to the users and the minimum number of users required for successful building up a database. In the second stage, each user decided whether to collaborate or not. Yang *et al.* [8] proposed two incentive mechanisms for the platform-centric model and the user-centric model, respectively. For platform-centric model, only one sensing task was considered and a Stackelberg game was adopted, where the SP announced a total reward and users determined their own sensing plan accordingly. This mechanism ensures that the utility of platform is maximized but is not truthful. For user-centric model, multiple sensing tasks were considered and a reverse auction-based method was proposed. Users can select several tasks and submit their claimed bid price to the platform, then the platform will determine the winning users. Koutsopoulos [9] considered the reverse auction design in a participatory sensor network. First, the customers sent queries to the service provider for the requested data. Next the SP sent a request for data contribution to the mobile users. The mobile users then declared their participation costs, which was private information, to the SP. An incentive-compatible mechanism was proposed to specify the participation level and payment of each mobile user. The objective of the SP was to minimize the total payment to the MUs, subject to the quality of experience of the customers. Feng *et al.* [10] took the location information into consideration and designed a truthful auction scheme. Jaimes *et al.* [11] also considered the location information of smartphone users when selecting the participants. It considered the coverage and the budget constraint of SP at the same time. Chen *et al.* [12] used the double auction to design the incentive mechanism for both multiple sensing tasks and smartphone users.

Most of related works on incentive mechanism design focus on the participation of smartphone users, in other words, the data quantity. However, data quality is also an important factor for a data collector since smartphone users may submit some erroneous or unreliable data. Low-quality data will impact the accuracy of data analysis result and degrade the provided service. Therefore, data quality should be considered in the incentive mechanism design. Reputation is one way to evaluate the quality of data provided by a mobile user. Kantarci and Mouftah [13] considered the reputation in the collection of sensing data. However, the proposed method mainly focused on the improvement of SP utility of SP and cannot guarantee the truthfulness.

Therefore, we aim to take the data quality into the consideration of designing efficient incentive mechanism with the property of truthfulness and individual rationality. We use the reputation to evaluate the quality of sensing data a smartphone user submits. The reputation of each smartphone user can be updated using a reputation system like the one proposed in Reference 9. It reflects the accuracy of recent reported data. With the consideration of the reputation of smartphone users, we design a truthful incentive mechanism which can maximize the weighted social welfare, and improve the overall quality of collected data as well.

7.2 System model

As in Figure 7.1, we consider a participatory sensing system consisting of an SP and multiple smartphone users. The SP is the promulgator of sensing tasks. It aims to collect the information in an area through the involvement of mobile users. SP posts the tasks through mobile phone apps. The smartphone users are the executors of sensing tasks. They can download and install these apps, and participate in the sensing procedure. Being at different locations in this area, the smartphone users can execute these sensing tasks, and submit their sensing data to the SP. The whole sensing area is divided into many grids denoted by the set $G = \{1, 2, \ldots, m\}$ and one user's sensing area can cover multiple grids, as the dashed circle illustrated in Figure 7.1.

Considering that locations may have different degrees of importance to SP, we let v_g represent the significance of the grid g, which denotes the valuation SP can obtain if she gets the sensing data of the grid g. Let $N = \{1, 2, \ldots, n\}$ be the set of smartphone users who are interested in the sensing task. Any user $i \in N$ is associated with several attributes:

- The set $G_i \subseteq G$: which represents the set of grids covered by the user i.
- The true cost c_i: which reflects the real cost for the user i to conduct the sensing task and report the sensing data. Note that true cost c_i is the private information for the user i and not open to any other users and SP.
- The bid price b_i: which is the payment that user i hope to receive from the SP to compensate the cost of conducting sensing task such as energy consumption and transmission cost. Note that the bid price b_i may not equal to the true cost c_i. For instance, the user i may request a bid price higher than its true cost if it can make her achieve a higher payment.
- The reputation r_i: which reflects the quality of sensing data provided by the user i. Base on the history of the received data, SP evaluates the reputation of each user. Note that the SP evaluates, updates, and keeps the reputation of each smartphone user.

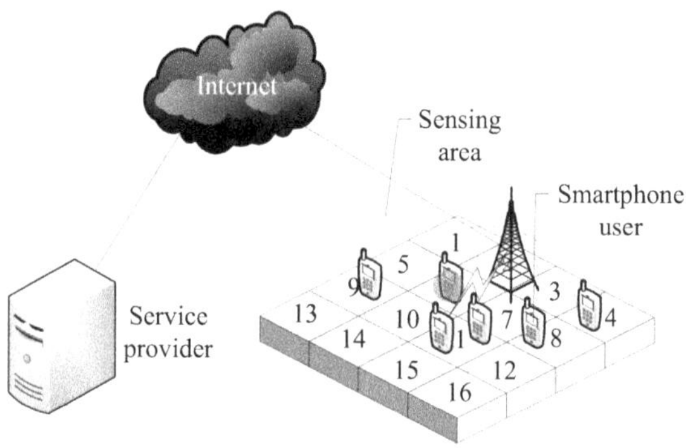

Figure 7.1 The illustration of a participatory sensing network

We take the quality of sensing data (i.e., the reputation of smartphone users) into consideration and use the auction method to design an incentive mechanism. Smartphone users interested in conducting the sensing task submit their own bid price b_i to SP. Based on the collected bid price $b_i, i \in N$, the reputation of each user r_i and the values of grids that each user can cover, SP determines the set of smartphone users to conduct the sensing task, and calculate the payment for each user. Then, winning users execute the sensing task and submit their sensing data to SP, and receive corresponding payment from SP. The allocation rule (winner selection) and payment rule of the designed incentive mechanism are presented in the following section.

7.3 Problem formulation

Usually, the primary objective of a non-commercial SP (e.g., governmental department, a non-profit organization) is to maximize the social welfare of the system. In this chapter, due to the consideration of quality of sensing data (i.e., the reputation of smartphone users), we introduce the weighted social welfare, where the achieved value for the grids covered by the sensing data is weighted by the quality of the sensing data (i.e., the reputation). The weighted social welfare is defined as follows.

Definition 1 (Weighted social welfare). *Weighted social welfare of the participatory sensing system is defined as the total weighted valuation of grids covered by the sensing data that SP collects minus the total cost of smartphone users conducting the sensing task, which is given as:*

$$S = \sum_{g \in G} (v_g \times r_g) - \sum_{i \in N} (b_i \times a_i) \tag{7.1}$$

where $a_i = 1$ denotes user i is selected to perform the sensing task, $a_i = 0$ otherwise. $r_g = \max\{r_h : h \in \Gamma_g\}$ represents the highest quality (i.e., reputation) among all collected data which cover the grid g, and $\Gamma_g = \{i : i \in N, a_i = 1, d_i^g = 1\}$ denotes the set of all smartphone users who cover the grid g and are selected to conduct the sensing task. Similarly, $d_i^g = 1$ denotes that the grid g is covered by the sensing area of the user i, otherwise, $d_i^g = 0$. The frequently used notations in this chapter are listed in Table 7.1.

Definition 2 (Utility of smartphone user). *The utility of user $i \in N$ is defined as:*

$$U_i = \begin{cases} p_i - c_i & \text{if } a_i = 1 \\ 0 & \text{otherwise} \end{cases} \tag{7.2}$$

where p_i is the payment that user i receives from SP, and c_i is the true cost of user i.

Our objective is to design a truthful incentive mechanism denoted by $\Psi = (N, G, b, r, v)$, where given N, G, b, r and v, the SP determines the set of users to perform sensing task with the objective of maximizing the weighted social welfare in (7.1). Meanwhile, the SP calculates corresponding payment p_i for each user.

Table 7.1 Frequently used notations

Notation	Explanation
N	set of smartphone users
W	set of selected users
W_{-i}	set of selected users except i
G	set of grids
Π	feasible set of W
i,g	smartphone user, grid
S	weighted social welfare
G_i	the set of grids covered by user i
S_{-i}	weighted social welfare excluding user i from the system
S^*	optimal weighted social welfare
$S(W^*)$	optimal weighted social welfare with optimal winner W^*
Γ_g	subset of selected users who cover grid g
τ_i	the contribution set of user i
v_g	value of grid g
b_i	bid price of user i
c_i	true cost of user i
r_i	reputation of user i
p_i	payment to user i
U_i	utility of user i
b	bid vector
r	reputation vector
v	grid value vector
n, m	number of smartphone users, number of grids

We adopt the weighted Vickrey-Clarke-Groves (VCG) scheme to design a reputation-aware incentive mechanism (RAIM). The winner selection rule (i.e., allocation rule) and payment rule of the proposed RAIM are elaborated in detail as follows.

7.3.1 Allocation rule

Based on the collected bid prices, reputation, sensing coverage of each user, and the significance of each grid (i.e., the value of each grid), the SP selects a set of users to perform the sensing task with the objective of maximizing the weighted social welfare, which can be formulated as:

$$W^* = \arg\max_{W \in \Pi} S = \sum_{g \in G} (v_g \times r_g(W)) - \sum_{i \in N} (b_i \times a_i(W)) \tag{7.3}$$

where W denotes a possible allocation result (i.e., the set of users selected to perform the sensing task). Π is the set of all possible allocation results that SP can adopt, and $r_g(W) = \max\{r_h : a_h(W) = 1, d_h^g = 1, h \in N\}$, $a_i(W) = 1$ if $i \in W$.

Algorithm 1 Proposed incentive mechanism (N, G, b, r, v)

//Stage 1: Winner Allocation
$S^* = \max S$
$W^* = \arg\max_{W \in \Pi} S$
//Stage 2: Payment calculation
for all $i \in N$ **do**
  $p_i = 0$
end for
for all $i \in W^*$ **do**
  $S_{-i} = \max S_{N \setminus \{i\}}$
  $p_i = b_i + (S^* - S_{-i})$
end for
return (S^*, W^*, p)

7.3.2 Payment rule

The SP determines the payment p_i for each user i. If user i is not selected to perform the sensing task (i.e., $a_i = 0$), the payment $p_i = 0$; If user i is a winner (i.e., $a_i = 1$), the payment is

$$\begin{aligned} p_i &= b_i + (S^* - S_{-i}) \\ &= \sum_{g \in G} (v_g \times r_g(W^*)) - \sum_{j \in N, j \neq i} (b_j \times a_j(W^*)) - S_{-i} \end{aligned} \tag{7.4}$$

where S_{-i} denotes the optimal weighted social welfare when excluding user i from the auction. $r_g(W^*) = \max\{r_h : a_h(W^*) = 1, d_h^g = 1, h \in N\}$, $a_j(W^*) = 1$ if $j \in W^*$.

7.3.3 Proof of properties

In this subsection, we prove our proposed incentive mechanism is individually rational and truthful.

Theorem 1 (Individually rational). *Our proposed incentive mechanism is individually rational.*

Proof. For the users who are not selected as winners, their utility is zero. For each winning user $i \in W^*$, his utility U_i satisfies the following property

$$\begin{aligned} U_i &= p_i - c_i \\ &= b_i + (S^* - S_{-i}) - c_i \end{aligned} \tag{7.5}$$

When bidding truthfully, i.e., $b_i = c_i$, $U_i = S^* - S_{-i} \geq 0$. So our proposed incentive mechanism achieves individually rational. □

Theorem 2 (Truthfulness). *Our proposed incentive mechanism is truthful.*

Before we prove this theorem, we first prove several lemmas.

Definition 3 (Contribution set of user i). *Given user i is selected to conduct sensing task, the contribution set of user i, denoted as τ_i, is defined as the set of grids that user i cover and has the highest reputation over all other selected users covering these grids. We have $\tau_i = \{g : g \in G, r_i = \max\{r_h : h \in \Gamma_g\}\}$.*

Let $W = \{i\} \bigcup W_{-i}$ be the set of all selected smartphone users, where W_{-i} denotes the set of all selected users except i.

Lemma 1. *Given the selection result $W = \{i\} \bigcup W_{-i}$ maximize the weighted social welfare defined in (7.3), then after excluding user i and removing the grids in the contribution set τ_i, the users in the set W_{-i} maximize the weighted social welfare of the remaining system.*

Proof. Given the winner set $W = \{i\} \bigcup W_{-i}$. Based on the definition of weighted social welfare, user i has no contribution to the grids except τ_i, and all the other users have no contribution to the grids in τ_i. So, when we remove i and τ_i from the system, the remaining system is not affected. So, the set of users in W_{-i} can still maximize the weighted social welfare of the remaining system. Hence, Lemma 1 holds. □

Lemma 2. *If user i wins by bidding b_i, she can also win by bidding $b_i' < b_i$.*

Proof. The selection rule is to maximize the weighted social welfare, so when user i wins the auction with the bid price b_i, $i \in W^*$, we can get the largest weighted social welfare S^*. When user i bids $b_i' < b_i$, based on (7.3), we will get a larger weighted social welfare S' (i.e., $S' > S^*$). So, user i must be a winner too when bidding $b_i' < b_i$. □

Based on the above lemmas, we prove Theorem 2 by showing that no user can improve her utility by bidding different from her true cost, i.e., $U_i \geq U_i'$ for any $b_i \neq c_i$, where U_i and U_i' are the utility of user i when bidding c_i and b_i, respectively. We verify all possible cases to show the truthfulness.

Case 1: $b_i = c_i$
Based on the payment rule of our proposed mechanism in (7.4), we can easily see that if $b_i = c_i$, the payment $p_i' = p_i$, so we have $U_i' = p_i' - c_i = p_i - c_i = U_i$.

Case 2: $b_i > c_i$
Based on Lemma 3, the case user i loses by bidding c_i but wins by bidding b_i is not impossible. So we need to consider three subcases:

Subcase 1: User i wins by bidding both b_i and c_i. In this case, assume the winner set is denoted as W^* when user i bidding c_i. Let $W^* = \{i\} \bigcup W^*_{-i}$ be the selection result. Then user i's utility is

$$
\begin{aligned}
U_i &= p_i - c_i \\
&= c_i + (S(W^*) - S_{-i}) - c_i \\
&= \sum_{g \in G} (v_g \times r_g(W^*)) - \sum_{j \in N} (c_j \times a_j(W^*)) - S_{-i}
\end{aligned} \tag{7.6}
$$

where $r_g(W^*) = \max\{r_h : a_h(W^*) = 1, c_h^g = 1, h \in N\}$, and $a_j(W^*) = 1$ if $j \in W^*$.

Assume the winner set is W' when user i bid with b_i, let $W' = \{i\} \bigcup W'_{-i}$ be the selection result. Similarly, user i's utility is

$$
\begin{aligned}
U_i' &= p_i' - c_i \\
&= b_i + (S(W') - S_{-i}) - c_i \\
&= \sum_{g \in G} (v_g \times r_g(W')) - \sum_{j \in N} (c_j \times a_j(W')) - S_{-i}
\end{aligned} \tag{7.7}
$$

Based on Lemma 1, if the winner set is $W^* = \{i\} \bigcup W^*_{-i}$, then the subset W^*_{-i} maximize the weighted social welfare of the remaining system, which actually is excluding user i and removing the grids in τ_i. We assume the whole system is divided into two parts: τ_i and G_{-i}, where $G_{-i} = G \setminus \tau_i$. For the part τ_i, i can maximize it. For the part G_{-i}, W'_{-i} is also the winner set. So the winner set W^* and W' are same. Hence, based on (7.6) and (7.7), $U_i' = U_i$.

Subcase 2: User i wins by bidding c_i but loses by bidding b_i. In this case, it is clear that $U_i' = 0$. $U_i = p_i - c_i = b_i + (S^* - S_{-i}) - c_i$, because $S^* \geq S_{-i}$ and $b_i > c_i$, so $U_i \geq 0$. Hence, $U_i' \leq U_i$.

Subcase 3: User i loses by bidding both b_i and c_i. In this case, it is clear that $U_i' = U_i = 0$.

Case 3: $b_i < c_i$
Based on Lemma 3, the case user i wins by bidding c_i but lose by bidding b_i is impossible. So there are three subcases too.

Subcase 1: User i wins by bidding both b_i and c_i. We can follow the same analysis in case 2 above to prove $U_i' = U_i$.

Subcase 2: User i loses by bidding c_i but wins by bidding b_i. In this case, $U_i = 0$. Assume the optimal weighted social welfare is $S(W^*)$ and $S(W')$ when user i bids c_i and b_i, respectively. Based on (7.4), when user i bids b_i and wins the auction, his/her payment $p_i = b_i + (S(W') - S_{-i})$. As we know S_{-i} is the optimal social welfare when we exclude user i in the auction, actually, its result is equivalent to the case when user i loses. So $S_{-i} = S(W^*)$. We can calculate the weighted social welfare (not optimal) when user i bids c_i with the strategy W', we denote this value as s, based on the fact W' is not the optimal winner set when bidding c_i but is the optimal winner set when bidding b_i, $S(W') = s + (c_i - b_i)$. Hence:

$$\begin{aligned} U_i' &= p_i - c_i \\ &= b_i + (S(W') - S_{-i}) - c_i \\ &= b_i + (s + (c_i - b_i) - S(W^*)) - c_i \\ &= s - S(W^*) \end{aligned} \tag{7.8}$$

Because $S(W^*) \geq s$, we get $U_i' \leq 0 = U_i$.

Subcase 3: User i loses by bidding both b_i and c_i. In this case, $U_i' = U_i = 0$.

So, based on all the cases talked above, we can get $U_i \geq U_i'$. Hence our proposed incentive mechanism is truthful.

7.4 Performance evaluation

Simulations have been conducted to evaluate the performance of the proposed incentive mechanism (i.e., RAIM) in terms of the achieved weighted social welfare and the average reputation of users selected to conduct the sensing task. We compare RAIM with three other mechanisms: random selection, reverse auction, and TSCM proposed in Reference 13. For random selection, we randomly select smartphone users as winners and the total number of winning users is the same as that in RAIM. For reverse auction, we first sort all users' bids in ascending order, and then select the users with smaller bid as winners, and the number of winners is same as RAIM as well. In TSCM, the winner selection rule is based on users' reputable marginal value minus their modified bids. The modified bid of a user is his actual bid divided by his reputation score.

7.4.1 Simulation setup

We consider a participatory sensing system in which the whole sensing area is a 200 m × 200 m square. This square is divided into multiple square grids with 10 m × 10 m. Smartphone users are randomly located in this area, and each user's sensing coverage includes all the grids within a distance of 20 m from this user, as shown in Figure 7.2. The bid of each user is uniformly distributed over [1,10]. The reputation of each user is uniformly distributed over [0,1]. We consider two scenarios. In the first scenario, the value of each grid is uniformly distributed over [1,5]. In the second scenario, there are three hotspots, in which the grids close to the center of these hotspots are much more important to the SP, so they have a larger value,

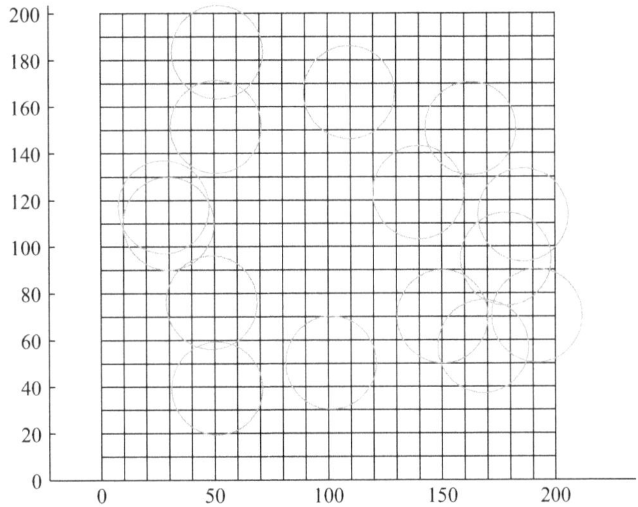

Figure 7.2 Simulation setup: the circle denotes user's sensing area

as shown in Figure 7.3. We set the average value of grids in these two scenarios be the same.

7.4.2 Truthfulness

To verify the truthfulness of the proposed incentive mechanism, we randomly pick up one user (ID = 1) whose true cost is 8 and observe how his utility changes when he bids different values from 1 to 20. Figure 7.4 shows the property of truthfulness since the user cannot improve her utility by bidding untruthfully.

7.4.3 Weighted social welfare

Figure 7.5 plots the weighted social welfare with different numbers of grids. Compared with the random selection and reverse auction, the proposed mechanism RAIM can achieve the largest weighted social welfare. The trends for both scenarios are the same. Due to the space limit, we only present the simulation results in Scenario 1 here.

Figure 7.6 shows the impact of number of users on weighted social welfare. It is observed that the proposed RAIM outperforms other counterparts. Compared with TSCM and random selection, RAIM improves the weighted social welfare by 8.65% and 48.16%, respectively, with the number of users N = 18.

7.4.4 Average reputation

We use the average reputation over smartphone users selected to conduct the sensing task to reflect the quality of collected sensing data. From Figure 7.7, it is observed that RAIM can achieve the largest average reputation, which means that the proposed mechanism can achieve a higher quality sensing data.

7.5 Conclusion and discussion

We have designed the incentive mechanism RAIM to maximize the weighted social welfare and improve the quality of sensing data. Our proposed incentive mechanism is truthful and individually rational. The simulation results have demonstrated the

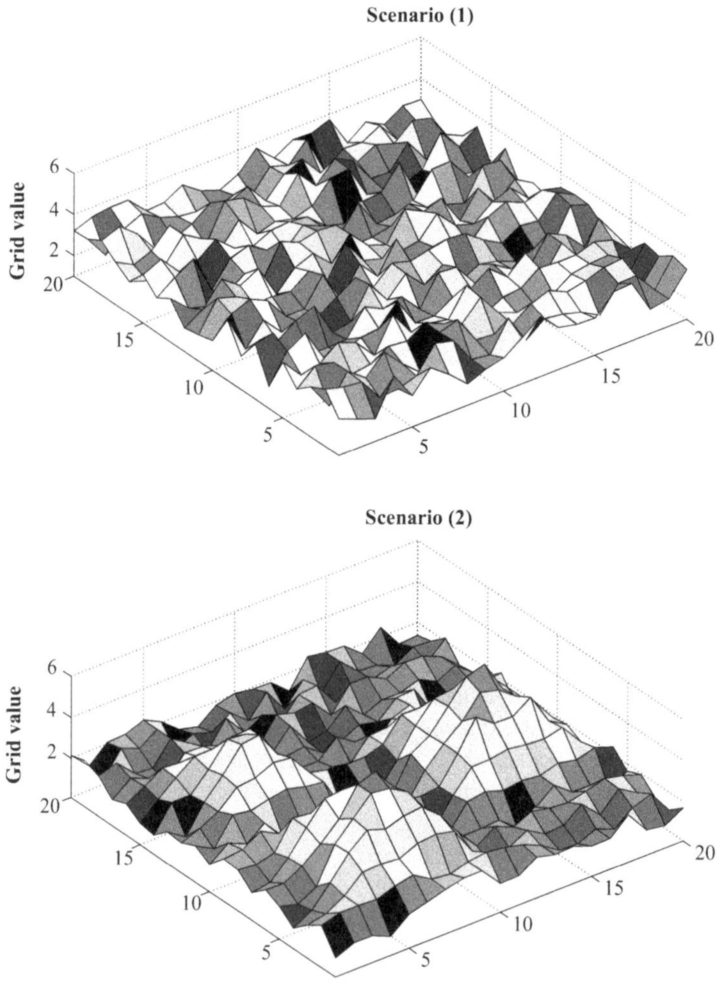

Figure 7.3 Scenario (1) no hotspot; Scenario (2): three hotspots

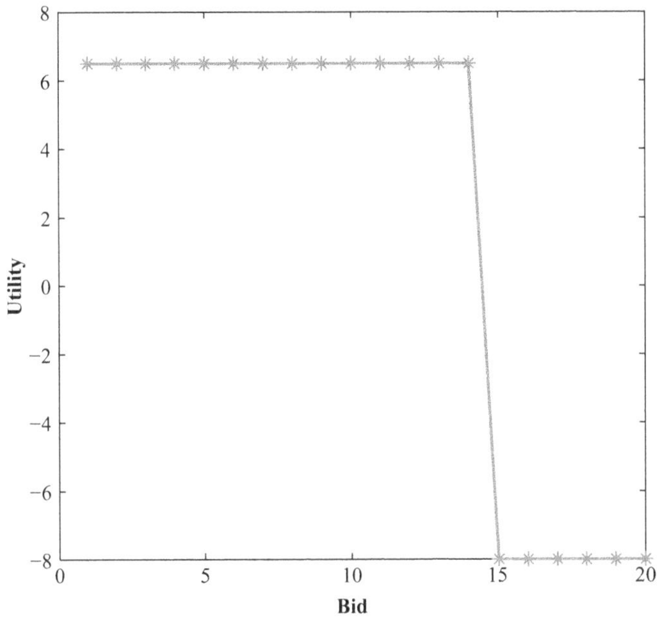

Figure 7.4 $c_l = 8$

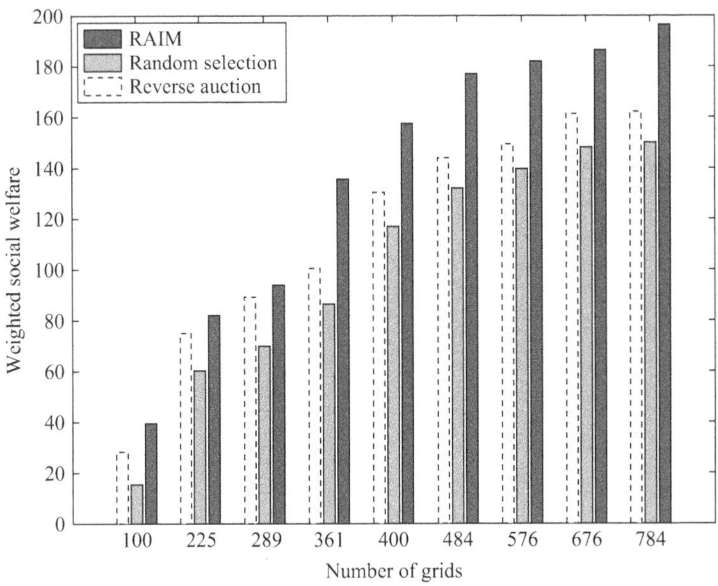

Figure 7.5 Weighted social welfare with different number of grids in scenario 1, n = 15

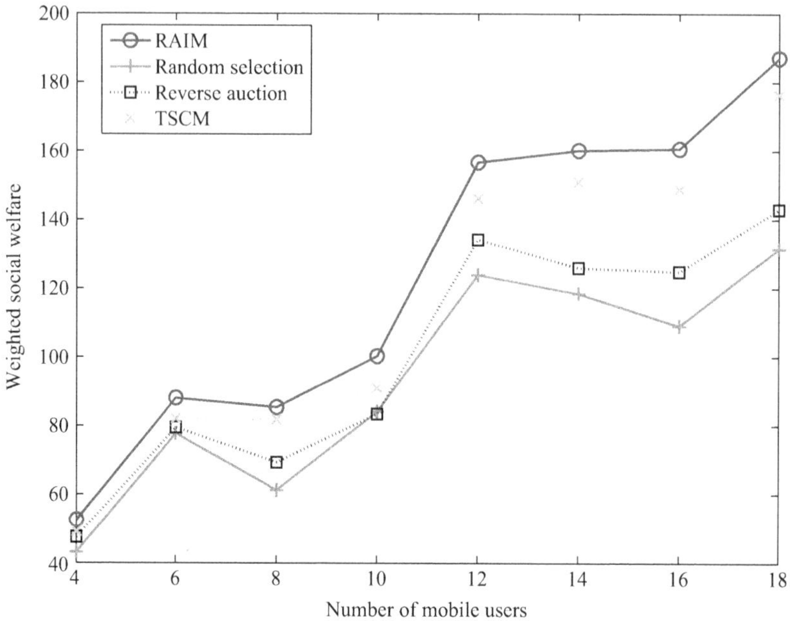

Figure 7.6 Impact of number of users on weighted social welfare, m = 400

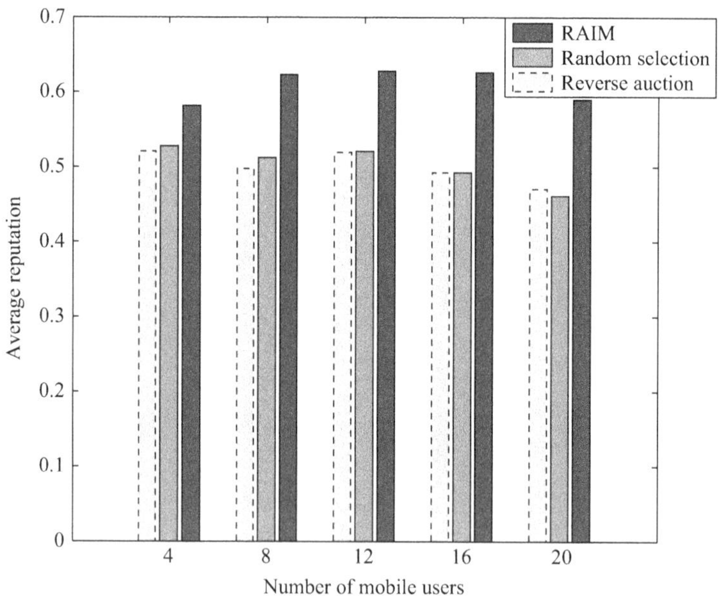

Figure 7.7 Impact of number of users on average reputation, m = 400

better performance of our proposed mechanism in terms of weighted social welfare, average reputation, and truthfulness. In addition to the incentive mechanism design, another interesting research issue is the integration of PSN with online social network, which is driven by the popularity of online social network (e.g., Facebook, WeChat, Twitters, QQ) and benefit both networks.

Acknowledgements

This work is supported by the Research Committee of University of Macau under grants MYRG2016-00171-FST, MYRG2014-00140-FST, MYRG101(Y1-L3)-FST13-MSD and MYRG2014-00146-FST, and by the Macau Science and Technology Development Fund under grants FDCT 121/2014/A3 and FDCT 091/2015/A3.

References

[1] Atzori L., Iera A., and Morabito G. 'The Internet of Things: a survey'. *Computer Networks*, vol. 54, no. (15), pp. 2787–2805, 2010.

[2] Sha K., Zhan G., Shi W., Lumley M., Wiholm C., and Arnetz B. 'Spa: a smart phone assisted chronic illness self-management system with participatory sensing'. *Proceedings of ACM HealthNet*; Breckenridge, USA, June 2008, pp. 1–5.

[3] Ericsson J. 'Ericsson mobility report'. 2014. Available from http://www.ericsson.com/mobility-report

[4] Lane N.D., Miluzzo E., Lu H., Peebles D., Choudhury T., and Campbell A.T. 'A survey of mobile phone sensing'. *IEEE Communications Magazine*, vol. 48, no. 9, pp. 140–150, 2010.

[5] Ganti R.K., Ye F., and Lei H. 'Mobile crowdsensing: current state and future challenges'. *IEEE Communications Magazine*, vol. 49, no. 11, pp. 32–39, 2011.

[6] Lee J.S., and Hoh B. 'Sell your experiences: a market mechanism based incentive for participatory sensing'. *Proceedings of IEEE International Conference on Pervasive Computing and Communications (PerCom)*; Mannheim, Germany, March 2010, pp. 60–68.

[7] Duan L., Kubo T., Sugiyama K., Huang J., Hasegawa T., and Walrand J. 'Incentive mechanisms for smartphone collaboration in data acquisition and distributed computing'. *Proceedings of IEEE INFOCOM*; Orlando, USA, March 2012, pp. 1701–1709.

[8] Yang D., Xue G., Fang X., and Tang J. 'Crowdsourcing to smartphones: incentive mechanism design for mobile phone sensing'. *Proceedings of ACM 18th annual international conference on Mobile computing and networking*; Istanbul, Turkey, Aug 2012, pp. 173–184.

[9] Koutsopoulos I, 'Optimal incentive-driven design of participatory sensing systems'. *Proceedings of IEEE INFOCOM*; Turin, Italy, April 2013, pp. 1402–1410.

[10] Feng Z., Zhu Y., Zhang Q., Ni L.M., and Vasilakos A.V. 'Trac: Truthful auction for location-aware collaborative sensing in mobile crowdsourcing'. *Proceedings of IEEE INFOCOM*; Toronto, Canada, April 2014, pp. 1231–1239.

[11] Jaimes L.G., Vergara-Laurens I., and Labrador M.A. 'A location-based incentive mechanism for participatory sensing systems with budget constraints'. *Proceedings of IEEE International Conference on Pervasive Computing and Communications (PerCom)*; Lugano, Switzerland, March 2012, pp. 103–108.

[12] Chen C. and Wang Y. 'Sparc: Strategy-proof double auction for mobile participatory sensing'. *Proceedings of IEEE International Conference on Cloud Computing and Big Data (CloudCom-Asia)*; Fuzhou, China, December 2013, pp. 133–140.

[13] Kantarci B. and Mouftah H.T. 'Reputation-based sensing-as-a-service for crowd management over the cloud'. *Proceedings of IEEE International Conference on Communications (ICC)*; Sydney, Australia, June 2014, pp. 3614–3619.

Chapter 8
Economics of Internet of Things (IoT): market structure analysis

Cheng Zhang[1]

Abstract

Over the last few years, the Internet of Things (IoT) has obtained huge attention from both industry and academia. IoT is a novel paradigm of connecting every possible things through the current Internet. The core idea is to integrate a variety of things or objects around us with the global Internet through wired or wireless networks. The "things" being equipped with sensors, and (or) actuators have powerful computation ability. The sensor-generated data are transmitted through the networks for further process, while the actuators can receive instructions from network to perform certain tasks. While most of the current research studies the technical aspects of IoT, we concentrate on the economical aspect of an IoT system. Except for some government-dominated nonprofit projects, most of the IoT systems seek to improve profit. Therefore, we study how to price the IoT service under different market structures to maximize the IoT service providers' revenue.

8.1 Introduction

Over the last few years, the Internet of Things (IoT) has obtained huge attention from both industry and academia. IoT is a novel paradigm of connecting every possible things through the current Internet. The core idea is to integrate a variety of things or objects around us with the global Internet through wired or wireless networks. The "things" being equipped with sensors, and (or) actuators have powerful computation ability. The sensor-generated data are transmitted through the networks for further process , while the actuators can receive instructions from network to perform certain tasks. For example, a Nursing home IoT system (see Figure 8.1) may compose objects with gyro sensors, and the objects are connected to the Internet through cellular wireless network or Wi-Fi access point. The objects can detect whether the elderly people fell down in the nursing home. Once the elderly people fell down, the information

[1]School of Fundamental Science and Engineering, Waseda University, Tokyo, Japan

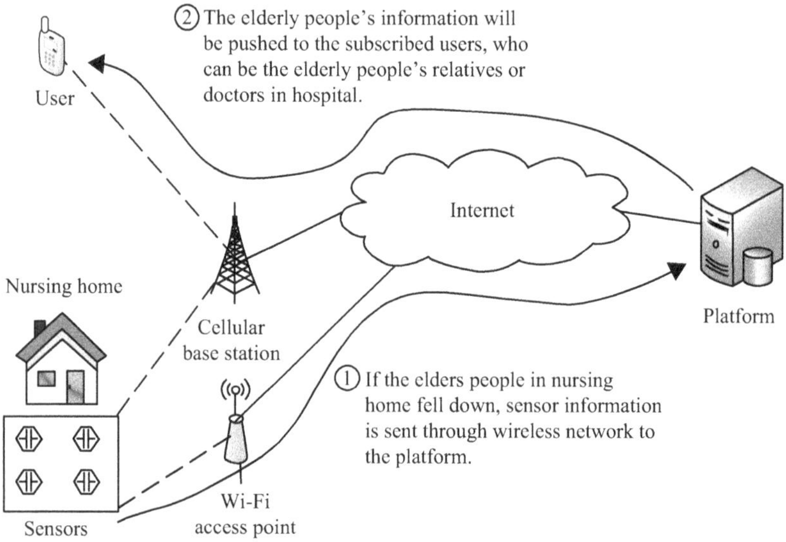

Figure 8.1 IoT solution for a nursing home

will be sent to the platform. On the other hand, the subscribed users who want to know the status of the elderly people will get the information from the platform through wired or wireless networks.

As shown in Figure 8.2, horizontally, IoT has the potential to improve system performance in multidisciplinary such as healthcare, transportation, manufacturing, and so on [1]. Vertically, the IoT system also have cross layers. We will introduce the layered IoT system from the bottom.

- ***Objects layer***: The objects layer is at the bottom, including low-level devices such as various kinds of sensors or actuators. Information is generated from this layer or the instructions are performed in this layer. Various network interfaces may be incorporated into the objects to support data transfer through heterogeneous network. They either direct connect to the communication layer through wireless networks, or connect among each other with peer-to-peer connections for data forwarding.
- ***Communication layer***: Data from/to objects is transferred through the communication layer. Wireless networks such as cellular wireless network, Wi-Fi access network, or WiMAX access network are used to connect the objects and the gateways, which are further connected to the Internet backbone. Data transmission with high reliability, high quality, and high security are always the main requirement for the communication layers. Even current telecommunication technologies can be directly applied to IoT, some special requirements of IoT have made how to design a suitable communication layer a challenging problem.
- ***Application layer***: This layer provides facility for data access and data storage. Data from objects layer are processed and stored in this layer. Based on the

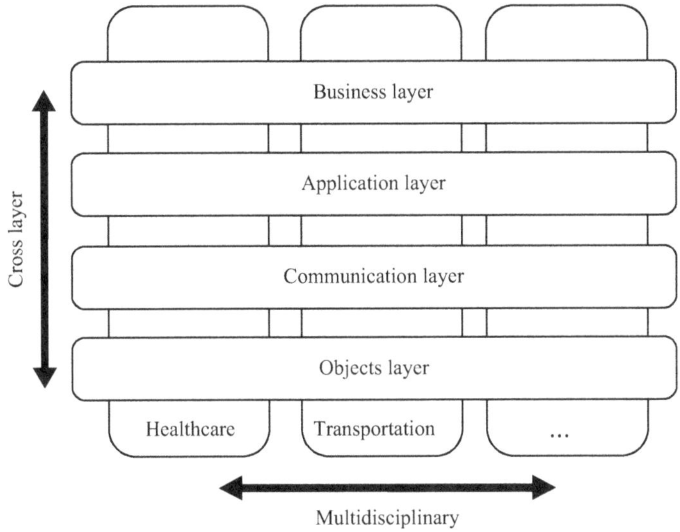

Figure 8.2 Layered IoT

collected data, services can be produced and then be provided for IoT users. It can also be combined with cloud system such as Infrastructure-as-a-Service (IaaS), Platform-as-a-Service (PaaS).

- ***Business layer***: This layer provides business-related functions for the IoT system, including business model, business echo system, price and cost, and so on. In order to provide sustainable IoT system, not only the technologies, but also the the aspect of business take an important role. Incentives should be provided for the players (such as network service providers, data service provider, and end users) involving the IoT system.

In this chapter, we concentrate on the business layer of an IoT system. Except for some government-dominated nonprofit projects, most of the IoT systems seek to improve profit. Therefore, we study how to price the IoT service under different market structures to maximize the IoT service providers profit.

In what follows, we first review some economics models of IoT. Then study the IoT service pricing problem for profit maximization under monopoly, duopoly, and oligopoly markets.

8.2 Economic models of IoT

For the business layer of IoT, there are still many aspects, this section discuss some economic models of IoT and related literature. We classify the economic models of IoT into three categories: *market model*, *utility model*, and *cost–benefit analysis model*.

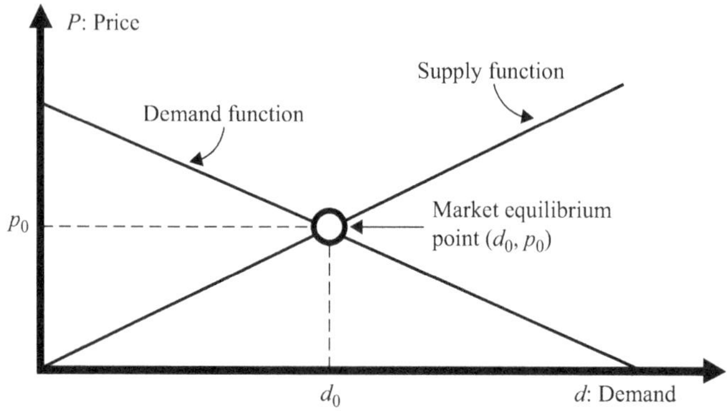

Figure 8.3 A simple market and pricing model

Market Model: Markets are where goods and services can be exchanged. Although money is not necessarily for exchange, it is always used for sellers and buyers to exchange their goods and services. Therefore, pricing has become an important factor for the market mechanism. As shown in Figure 8.3, demand for goods or services always decreases with the price, which is called demand function, and the supply for goods or services always increases with price, which is called supply function. The intersection of demand function curve and supply function curve is called market equilibrium, where the demand d_0 is fulfilled at the price p_0. There are different structures for a market.

The price determination depends on the specific market structure. For example, in a monopoly market, where only one seller dominated the market, pricing can be rather high to get the whole surplus of the buyers, while in a duopoly or oligopoly market, there are many sellers, who compete with each other, and arbitrary price is not possible since they should consider competitors pricing strategies, or the seller with high prices will lose buyers.

In IoT area, Munjin and Morin [2] draw an analogy between the smartphone application marketplaces, e.g., Apple AppStore and Google Play, and the existing similar trends in the IoT, and they propose a similar market for IoT for data and service exchange. However, the authors [2] did not consider the pricing problem in the proposed IoT marketplace.

Utility model: In economics, utility measures the satisfaction or preferences over some goods and services. The concept of utility has been adopted in telecommunication for bandwidth resource allocation. As shown in Figure 8.4, utility is a logarithmic function of allocated bandwidth, which has the property of "diminishing returns." Instead of allocating bandwidth equally for users, the bandwidth is allocated based on the utility of each user to maximize the overall satisfaction of all users.

Al-Fagih *et al.* [3] introduced utility to quantify the quality of service performance of the sensor data for a smart city system. The systems include access points,

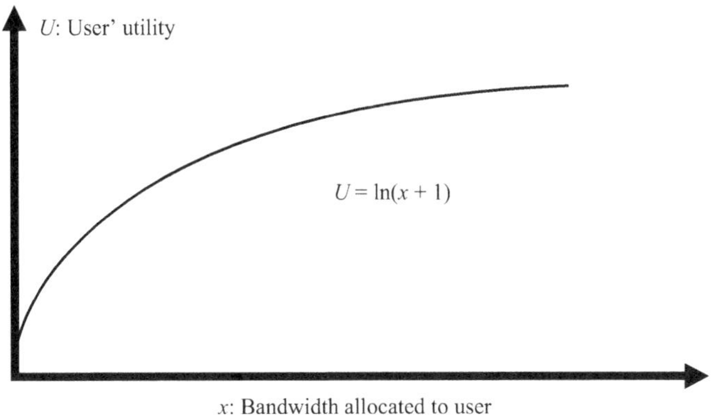

Figure 8.4 A kind of user's utility model

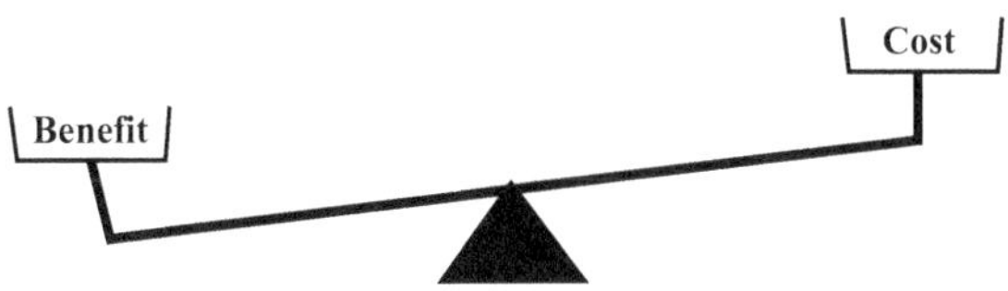

Figure 8.5 Cost–benefit analysis

gateways, data collectors, and sensor nodes. The access points initiate data request based on clients' request, and the date generated by the sensors are collected by data collectors then transferred by gateway. The utility is defined as a function of delay, quality, and trust parameters. The system tries to maximize the clients' gain through the aforementioned utility.

Cost–benefit analysis model: Cost–benefit Analysis (CBA) [4] is a systematic approach to estimating the strengths and weaknesses of alternatives that satisfy transactions, activities or functional requirements for a business. It is a technique that is used to determine options that provide the best approach for the adoption and practice in terms of benefits in labor, time and cost savings, etc. When the benefit is greater than cost (see Figure 8.5), a business will be operated.

In IoT area, Uckelmann [5] proposed an alternative approach to performance measurement and CBA for an RFID IoT in logistic application. They have found that the benefit could be distributed among different parties, and they also proposed a Cost and Benefit Sharing scheme.

In this chapter, we introduce the market model of IoT. Although there may be many parties involved in the IoT market, we consider two parties in the market: IoT service provider and users. The IoT service provider provides IoT service to users through IoT service platform (see Figure 8.6).

We study how to pricing the IoT services to maximize the profit of IoT service providers under monopoly, duopoly, and oligopoly IoT service markets.

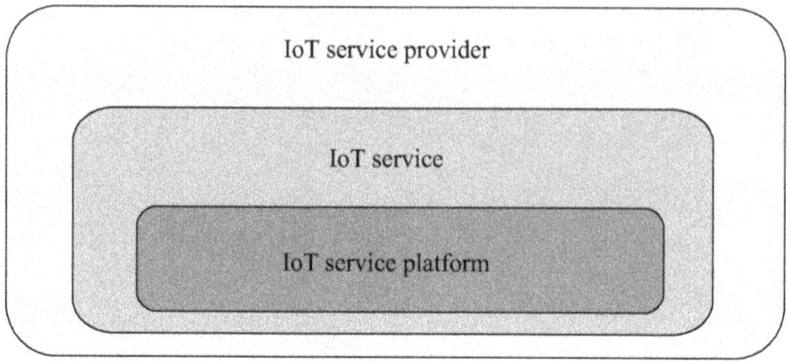

Figure 8.6 IoT service provider provides IoT service to users through IoT service platform

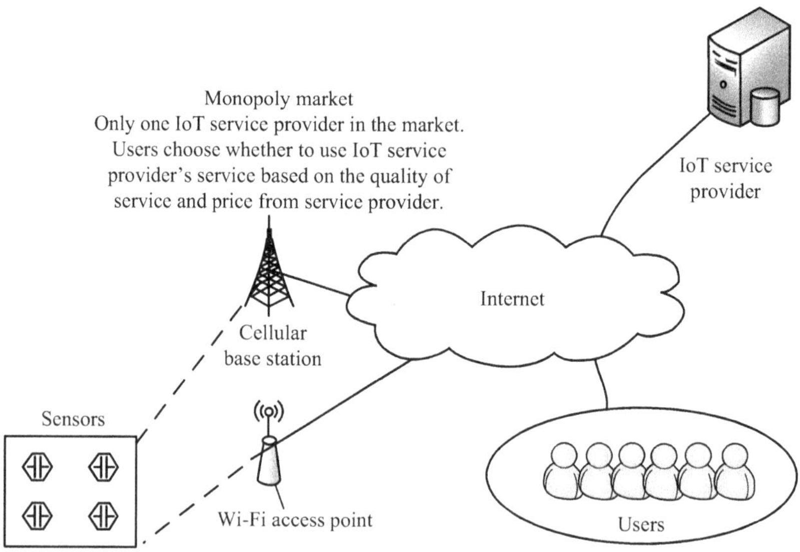

Figure 8.7 Monopoly market model

8.3 Monopoly market structure analysis of IoT

Under monopoly market, there is only one IoT service provider which denoted as $\mathbb{I}_1$, in the market (refer to Figure 8.7). $\mathbb{I}_1$ provides IoT services to users. The price set by $\mathbb{I}_1$ is p_1. It is assumed that the population of users denoted by M is fixed, with M_1 as the number of users using IoT service provider $\mathbb{I}_1$'s service. The proportion of users choosing $\mathbb{I}_1$'s service is denoted by (8.1).

$$x_1 = \frac{M_1}{M} \tag{8.1}$$

It is assumed that the value of M is very large.

The following set D defined in (8.2) is the domain for x_1.

$$D = \{x_1 | 0 \leq x_1 \leq 1\} \tag{8.2}$$

The quality of service provided by $\mathbb{I}_1$ is denoted as ϕ_1.

8.3.1 Monopoly market model

A continuum model of users is employed. If there are a large number of users in the market and each individual user is negligible, the continuum model approximates well the real user population. The payoff of user k is denoted as (8.3)

$$U_{k,1} = \lambda_k \phi_1 - p_1 \tag{8.3}$$

where $\lambda_k \in [0, 1]$ is user k's valuation for $\mathbb{I}_1$'s quality of service. The value of λ_k is private information to users, but the distribution of λ_k is public information of IoT service providers. Furthermore, different users may have different valuations on the same level of quality of service. ϕ_1 denotes the quality of service provided by $\mathbb{I}_1$ and $\lambda_k \phi_1$ denotes the benefit that the user can get from $\mathbb{I}_1$'s service. The unit of $\lambda_k \phi_1$ has the same unit as that of p_1, which is the price set by $\mathbb{I}_1$. Note that λ_k and ϕ_1 are normalized to [0, 1]. We have the following assumption for users' valuations of quality of service.

Assumption 1. *The users' valuations of quality of service have the probability density function (PDF)* $f(\cdot)$, *which is strictly positive and continuous on* [0, 1]. *The cumulative density function (CDF) is defined by* $F(a) = \int_{-\infty}^{a} f(y)\, dy$ *for all* $a \in R$.

Each user chooses to use $\mathbb{I}_1$'s service if he/she gets positive payoff by using $\mathbb{I}_1$. In other words, a user chooses an IoT service provider under the conditions enumerated as follows.

$$\begin{cases} \text{Use } \mathbb{I}_1\text{'s service} & \text{if } U_{k,1} > 0 \\ \text{Not use } \mathbb{I}_1\text{'s service} & \text{if } U_{k,1} \leq 0 \end{cases}$$

8.3.2 Monopoly market analysis

The monopoly IoT service provider tries to maximize profit. The profit of the IoT service provider depends on the number of users using his service, the service price and the cost, which is denoted as

$$\Pi_1 = p_1 x_1(p_1) - c_1(x_1) \tag{8.4}$$

Note that the number of user using service of $\mathbb{I}_1$, $x_1(p_1)$, is a function of the price set by $\mathbb{I}_1$, and liner cost function is assumed as shown in (8.5)

$$c_1(x_1) = c_1 x_1 \tag{8.5}$$

The parameter c_1 is also normalized to $[0, 1]$. The IoT service provider tries to maximize its profit by considering the following subproblem shown in (8.6):

$$\max_{p_1} \Pi_1 \quad \text{subject to } x_1 \in D \tag{8.6}$$

User k uses $\mathbb{I}_1$'s service if and only if the conditions shown in inequality (8.7) are satisfied.

$$\lambda_k \phi_1 - p_1 \geq 0 \tag{8.7}$$

Otherwise, user k does not use $\mathbb{I}_1$'s service.

Now we characterize the marginal points that identifying user's valuation of quality of service associated with changes in their decision to use IoT service provider $\mathbb{I}_1$'s service. $\tau^0_{\mathbb{I}_1}$ denotes the marginal point where users switch from getting negative payoff to deriving positive payoff from using IoT service provider $\mathbb{I}_1$, i.e., $\tau^0_{\mathbb{I}_1}$ is a point such that $U_{k,1} = 0$. We can have the following:

$$U_{k,1} = \lambda_k \phi_1 - p_1 > 0 \quad \text{if } \lambda_k > \tau^0_{\mathbb{I}_1} \tag{8.8}$$

Equation (8.8) indicates that if $\lambda_k > \tau^0_{\mathbb{I}_1}$, then the user with quality of service valuation greater than $\tau^0_{\mathbb{I}_1}$ can get positive payoff from using $\mathbb{I}_1$'s service. Therefore, it is very important to compute these marginal points, which determines whether the user uses $\mathbb{I}_1$'s service.

By setting $U_{k,i} = 0$, we can derive $\tau^0_{\mathbb{I}_1}$ shown in (8.9)

$$\tau^0_{\mathbb{I}_1} = \frac{p_1}{\phi_1} \tag{8.9}$$

Lemma 1. *The number of user who use $\mathbb{I}_1$'s service is shown as that in (8.10)*

$$x_1 = 1 - F(\tau^0_{\mathbb{I}_1}) \tag{8.10}$$

Proof. As shown in (8.8), if $\lambda_k > \tau^0_{\mathbb{I}_1}$, users will use IoT service provider $\mathbb{I}_1$'s network. Therefore, the number of users use IoT service provider $\mathbb{I}_1$'s network is as follows:

$$x_1 = P(\lambda_k > \tau^0_{\mathbb{I}_1}) \tag{8.11}$$

Equation (8.10) can easily be derived as follows.

$$x_1 = P(\lambda_k > \tau^0_{\mathbb{I}_1}) = 1 - P(\lambda_k \leq \tau^0_{\mathbb{I}_1}) = 1 - F(\tau^0_{\mathbb{I}_1}) \tag{8.12}$$

□

Lemma 1 shows that the number of users using $\mathbb{I}_1$'s service is a function of $\tau^0_{\mathbb{I}_1}$.

Proposition 1. *If users' quality of service valuation is distributed uniformly, then the optimal price for IoT service provider is as follows.*

$$p_1 = \frac{\phi_1 + c_1}{2} \tag{8.13}$$

Proof. Since $U_{k,1}$ is 0, when $\lambda_k = \tau^0_{\mathbb{I}_1}$. By the definition of $U_{k,1}$ in (8.3), the price set by IoT service provider $\mathbb{I}_1$ can be expressed as shown in (8.14).

$$p_1 = \tau^0_{\mathbb{I}_1}\phi_1 \tag{8.14}$$

Equation (8.14) shows that the price set by a monopoly is a function of $\tau^0_{\mathbb{I}_1}$.

If users' quality of service valuation is distributed uniformly, by Lemma 1, $x_1 = 1 - F(\tau^0_{\mathbb{I}_1}) = 1 - \tau^0_{\mathbb{I}_1}$.

Therefore, the maximization of Π_1 can be derived as shown in (8.15).

$$\begin{aligned}\max_{p_1} \Pi_1(p_1) &= \max_{p_1} (p_1 - c_1) * x_1 \\ &= \max_{\tau^0_{\mathbb{I}_1} \in [0,1]} [(\phi_1 \tau^0_{\mathbb{I}_1} - c_1) * (1 - \tau^0_{\mathbb{I}_1})]\end{aligned} \tag{8.15}$$

It is obvious that $\frac{d^2\Pi_1}{d(\tau^0_{\mathbb{I}_1})^2} = -2\phi_1 < 0$, and when $\tau^0_{\mathbb{I}_1} = \frac{\phi_1 + c_1}{2\phi_1} \in [0, 1]$, $\frac{d\Pi_1}{d\tau^0_{\mathbb{I}_1}} = 0$. Therefore, the solution for problem (8.15) is as in (8.16)

$$p_1 = \frac{\phi_1 + c_1}{2} \tag{8.16}$$

□

Proposition 1 shows that in a monopoly market, the IoT service provider sets price proportional to the sum of quality of service and the cost to maximize profit.

8.4 Oligopoly market structure analysis of IoT

Under oligopoly market, there are many competing IoT service providers. In this chapter, n competing IoT service providers are considered, which are denoted as $\mathbb{I}_1$, $\mathbb{I}_2$, ..., $\mathbb{I}_n$ respectively (refer to Figure 8.8). IoT service providers $\mathbb{I}_1$, $\mathbb{I}_2$, ..., $\mathbb{I}_n$ provide the substitute IoT services to users. IoT service provider $\mathbb{I}_i$, $i \in \{1, 2, \ldots, n\}$ sets price p_i (where $i \in \{1, 2, \ldots, n\}$). We denote that $\boldsymbol{p} = (p_1, \ldots, p_n)'$, where $\boldsymbol{p}$ is an $n \times 1$ column vector, and "$'$" means transpose operator. It is assumed that the population of users denoted by M is fixed, with M_i as the number of users choosing IoT service provider $\mathbb{I}_i$ for $i \in \{1, 2, \ldots, n\}$. The proportion of users who choose IoT service provider $\mathbb{I}_i$ is denoted by (8.17).

$$x_i = \frac{M_i}{M}, \quad \text{where } i \in \{1, 2, \ldots, n\} \tag{8.17}$$

It is assumed that the value of M is very large. We also denote that $\boldsymbol{x} = (x_1, x_2, \ldots, x_n)'$.

The following set D defined in (8.18) is the domain for $x_1, x_2, \ldots, x_n$:

$$D = \left\{ (x_1, x_2, \ldots, x_n) \middle| \sum_{i=1}^{n} x_i \leq 1, x_i \in [0, 1] \right\} \tag{8.18}$$

The level of quality of service provided by IoT service provider $\mathbb{I}_i$ for $i \in \{1, 2, \ldots, n\}$ is denoted as ϕ_i.

Assumption 2. *Without loss of generality, we assume that* $\phi_1 > \phi_2 > \cdots > \phi_n$.

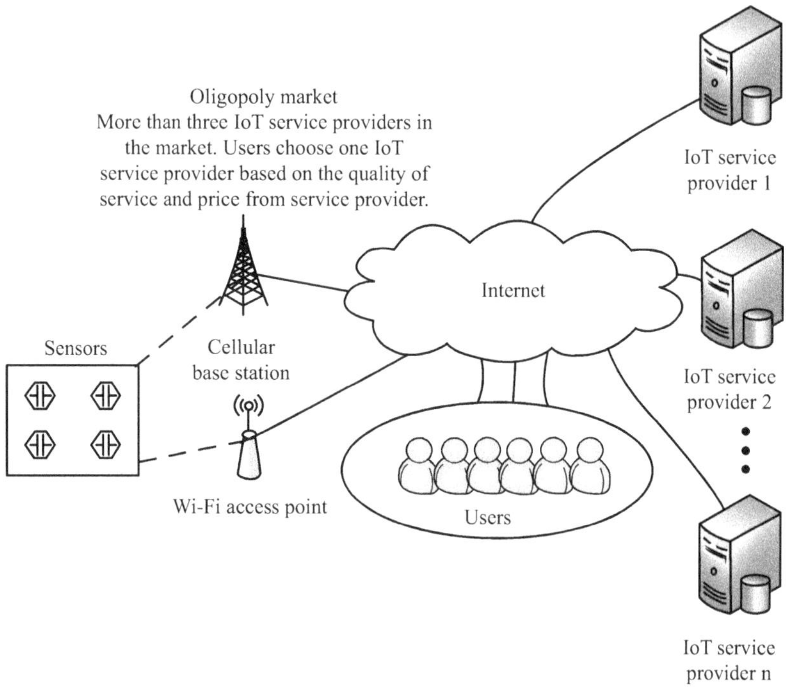

Figure 8.8 Oligopoly market model

8.4.1 Oligopoly market model

Similar to user model in Section 8.3.1, a continuum model of users is employed. The payoff of user k is denoted as (8.19)

$$U_{k,i} = \lambda_k \phi_i - p_i \tag{8.19}$$

where $\lambda_k \in [0, 1]$ is the quality of service valuation of user k. ϕ_i denotes the quality of service provided by IoT service provider $\mathbb{I}_i$, and $\lambda_k \phi_i$ denotes the benefit that the user can get from IoT service provider $\mathbb{I}_i$. The price set by IoT service provider $\mathbb{I}_i$ is denoted as p_i. Note that λ_k and ϕ_i are normalized to [0, 1].

For the users' valuations of quality of service, see the Assumption 1 in Section 8.3.1. We also assumed that there is no switching cost when users change from one IoT service provider to another. Each user makes decision independently.

8.4.2 Oligopoly market analysis

IoT service providers try to maximize their profit. We model the IoT service provider oligopoly competition as a Bertrand competition game. In the Bertrand competition game, different firms strategically choose prices independently at the same time while supplying quantities demanded at the chosen prices [6–8]. The profit of IoT service providers is defined as in (8.20).

$$\Pi_i = p_i x_i(\boldsymbol{p}) - c_i(x_i(\boldsymbol{p})) \tag{8.20}$$

$x_i(\boldsymbol{p})$ is the proportion of users who choose IoT service provider $\mathbb{I}_i$ as defined in (8.17). Note that it is a function of the prices set by all the network $\mathbb{I}_1$, $\mathbb{I}_2$, ..., $\mathbb{I}_n$, which can be calculated by Lemma 2 established in this section. $c_i(x_i(\boldsymbol{p}))$ is the linear cost function which is assumed as $c_i(x_i) = c_i x_i$.

User k chooses IoT service provider that gives him/her highest payoff. The index of IoT service provider that user k chooses can be expressed as shown in (8.21).

$$l = \arg \max_{i \in 1,\ldots,n} U_{k,i} \tag{8.21}$$

Now we characterize the marginal points that identify users' valuation of quality of service associated with changes in their decision to choose IoT service provider. We have the following Definition 1.

Definition 1. *Define n marginal points* $\tau_k^1, \tau_k^2, \ldots, \tau_k^n$, *as*

$$\tau_k^i = \frac{p_i - p_{i+1}}{\phi_i - \phi_{i+1}} \quad \text{if } i = 1, 2, \ldots, n-1 \tag{8.22}$$

$$\tau_k^n = \frac{p_n}{\phi_n} \tag{8.23}$$

We can derive these marginal points by letting $U_{k,i} = U_{k,i+1}$ for $i = 1, 2, \ldots, n-1$ and $U_{k,n} = 0$. Geometrically, for $i = 1, 2, \ldots, n-1$, τ_k^i is the horizontal axis value of the intersection point between line $U_{k,i}$ and $U_{k,i+1}$. τ_k^n is the horizontal axis value of the intersection point between horizontal axis and line $U_{k,n}$.

In order to illustrate the marginal points in Definition 1, we assume that there are three IoT service providers, which means that $n = 3$. User's payoff function $U_{k,i} = \lambda_k \phi_i - p_i$ is a linear function of user's quality of service valuation λ_k, and the slope of the payoff function is ϕ_i. By Assumption 2, we have $\phi_1 > \phi_2 > \phi_3$. The user's payoff function lines can be drawn in the same coordinate system with λ_k as the horizontal axis and user's payoff as the vertical axis (see Figure 8.9). It is shown that the prices set by IoT service providers are the y-intercepts in the coordinate system. The marginal point τ_k^1 can be derived by letting $U_{k,1} = U_{k,2}$, which is $\tau_k^1 = \frac{p_1 - p_2}{\phi_1 - \phi_2}$, and the payoff lines $U_{k,1}$ and $U_{k,2}$ intersect at point E. Similarly, the marginal point τ_k^2 can be derived by letting $U_{k,2} = U_{k,3}$, which is $\tau_k^2 = \frac{p_2 - p_3}{\phi_2 - \phi_3}$, and payoff lines $U_{k,2}$ and $U_{k,3}$ intersect at point C. The marginal point τ_k^3 can be derived by letting $U_{k,3} = 0$, which is $\tau_k^3 = \frac{p_3}{\phi_3}$. The payoff line τ_k^3 and horizontal axis intersect at point A.

Lemma 2. *The necessary and sufficient condition for positive number of users in each IoT service provider's network is shown in (8.24),*

$$1 > \tau_k^1 > \tau_k^2 > \cdots > \tau_k^n \tag{8.24}$$

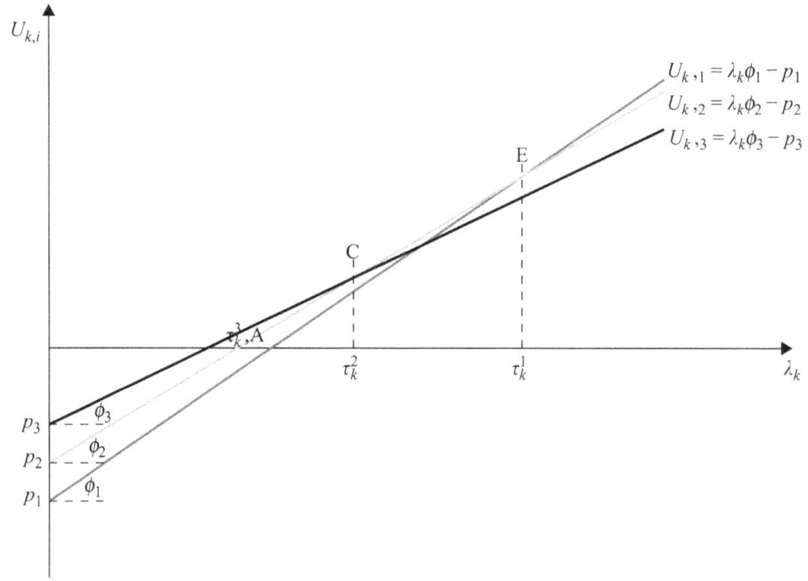

Figure 8.9 Illustration of marginal points of users valuation with three IoT service providers case

The number of users in each IoT service provider's network can be expressed as shown in (8.25).

$$x_i = \begin{cases} F(\tau_k^{i-1}) - F(\tau_k^i) & \text{if } i = 2, \ldots, n \\ 1 - F(\tau_k^1) & \text{otherwise} \end{cases} \tag{8.25}$$

Proof. (1) *Sufficiency:* If $1 > \tau_k^1 > \tau_k^2 > \cdots > \tau_k^n$, then the users with valuation of quality of service belonging to $[\tau_k^i, \tau_k^{i-1}]$ use $\mathbb{I}_i$'s IoT services at price p_i. (This is obvious by Definition 1, we will further illustrate this point by example after the proof of this lemma.) The number of users using $\mathbb{I}_i$'s IoT services is given by

$$x_i = \begin{cases} \int_{\tau_k^i}^{\tau_k^{i-1}} f(y)\mathrm{d}y = F(\tau_k^{i-1}) - F(\tau_k^i) & \text{if } i = 2, \ldots, n \\ \int_{\tau_k^1}^{1} f(y)\mathrm{d}y = 1 - F(\tau_k^1) & \text{otherwise} \end{cases} \tag{8.26}$$

According to Assumption 1, it is obvious that $x_i > 0$, which means that the number of users choosing $\mathbb{I}_i$'s IoT services is positive.

(2) *Necessity:* If the number of users of IoT service providers, or $x_i > 0$. By (8.26), if $i = 2, \ldots, n$, we have $F(\tau_k^{i-1}) - F(\tau_k^i) > 0$, or $F(\tau_k^{i-1}) > F(\tau_k^i)$. Since $F(\cdot)$ is an increasing function, we have (8.27)

$$\tau_k^{i-1} > \tau_k^i \tag{8.27}$$

If $i = 1$, we have $1 - F(\tau_k^1) > 0$. Since $1 = F(1)$, we have $F(1) - F(\tau_k^1) > 0$. Similarly, we have (8.28).

$$1 > \tau_k^1 \tag{8.28}$$

By combining (8.27) and (8.28), we can check that the condition in (8.24) is right. □

Although users' quality of service valuation is a random variable with CDF $F(\cdot)$, the proportion of users x_i calculated from $F(\cdot)$ is not a random variable. Once the $F(\cdot)$ and the marginal points are determined, the value of x_i is determined from (8.25).

In order to illustrate that a user with valuation of quality of service between $[\tau_k^i, \tau_k^{i-1}]$ using $\mathbb{I}_i$'s IoT service at price p_i in the proof process of Lemma 2, we take $n = 3$ as an example. In Figure 8.9, if the user's quality of service valuation is between $[\tau_k^3, \tau_k^2]$, the line $U_{k,3}$ is above lines $U_{k,1}$ and $U_{k,2}$, this means that the user can obtain maximum payoff with IoT service provider $\mathbb{I}_3$ in this case, then the user chooses $\mathbb{I}_3$. Similarly, if the user's quality of service valuation is between $[\tau_k^2, \tau_k^1]$, the line $U_{k,2}$ is above lines $U_{k,1}$ and $U_{k,3}$, then the user uses $\mathbb{I}_2$; if the user's quality of service valuation is between $[\tau_k^1, 1]$, the line $U_{k,1}$ is above lines $U_{k,2}$ and $U_{k,3}$, then the user uses $\mathbb{I}_1$.

Each IoT service provider tries to maximize its profit by considering the following problem shown in (8.29).

$$\begin{aligned} &\max_{p_i} \Pi_i \\ &\text{subject to } x_i \in D \end{aligned} \tag{8.29}$$

The problem in (8.29) can be solved by considering the game played by IoT service provider $\mathbb{I}_1, \mathbb{I}_2, \ldots, \mathbb{I}_n$. The Nash Equilibrium point is the solution of the problems. Now we consider that n IoT service providers play a Bertrand competition (or price competition) game. The Bertrand competition game Γ(**Player, Strategy, Payoff**) is described as follows:

- *Player*: The IoT service provider $\mathbb{I}_1, \ldots, \mathbb{I}_n$ are the n players in the game.
- *Strategy*: The strategy is the price set by IoT service provider $\mathbb{I}_i$ for $i \in \{1, 2, \ldots, n\}$.
- *Payoff*: The payoff is the profit gotten by IoT service provider $\mathbb{I}_i$ for $i \in \{1, 2, \ldots, n\}$.

In this game, IoT service providers $\mathbb{I}_1, \mathbb{I}_2, \ldots, \mathbb{I}_n$ set their price $p_1, p_2, \ldots, p_n$, respectively, to maximize their profit, which is different between the revenue and cost. The number of users in each IoT service provider's network can be derived by Lemma 2.

Proposition 2. *If users' quality of service valuation is distributed uniformly, the necessary and sufficient condition for unique Nash Equilibrium of the game* Γ*(**Player, Strategy, Payoff**) is as that in (8.30),*

$$|\boldsymbol{I} - \boldsymbol{A}| \neq 0 \tag{8.30}$$

where

$$\boldsymbol{A} = \begin{pmatrix} 0 & \frac{1}{2} & 0 & 0 & 0 & \dots & 0 \\ \alpha_1 & 0 & \beta_3 & 0 & 0 & \dots & 0 \\ 0 & \alpha_2 & 0 & \beta_4 & 0 & \dots & 0 \\ 0 & 0 & \alpha_3 & 0 & \beta_5 & \dots & 0 \\ \vdots & \vdots & \vdots & \ddots & \ddots & \ddots & \vdots \\ 0 & 0 & 0 & 0 & \alpha_{n-2} & 0 & \beta_n \\ 0 & 0 & 0 & 0 & 0 & \alpha_{n-1} & 0 \end{pmatrix}$$

$$\alpha_{i-1} = \begin{cases} \dfrac{\varphi(\phi_1 - \phi_2)}{2} & \text{if} \quad i = 1 \\ \dfrac{\phi_i - \phi_{i+1}}{2(\phi_{i-1} - \phi_{i+1})} & \text{else if } i = 2, \dots, n-1 \\ \dfrac{\phi_n}{2\phi_{n-1}} & \text{else} \quad i = n \end{cases} \tag{8.31}$$

$$\beta_{i+1} = \frac{\phi_{i-1} - \phi_i}{2(\phi_{i-1} - \phi_{i+1})} \quad \text{for } i = 2, \dots, n-1 \tag{8.32}$$

and $\boldsymbol{I}$ *is an* $n \times n$ *unit matrix with ones on the main diagonal and zeros elsewhere.*

And the Nash Equilibrium price is

$$p_i = \frac{|(\boldsymbol{I} - \boldsymbol{A})_i|}{|(\boldsymbol{I} - \boldsymbol{A})|} \tag{8.33}$$

where $(\boldsymbol{I} - \boldsymbol{A})_i$ *is the matrix formed by replacing the ith column of* $(\boldsymbol{I} - \boldsymbol{A})$ *by the column vector* μ^t. μ^t *is defined as* $\mu^t = (\alpha_0, 0, \dots, 0)'$. *The operator* $|\cdot|$ *on a matrix denotes the determinant [9] of the matrix. Thus,* $|(\boldsymbol{I} - \boldsymbol{A})|$ *is determinant of the matrix* $(\boldsymbol{I} - \boldsymbol{A})$, $|(\boldsymbol{I} - \boldsymbol{A})_i|$ *is determinant of the matrix* $(\boldsymbol{I} - \boldsymbol{A})_i$.

Proof. Refer to Appendix for the details of proof.

8.5 Conclusions

We study how to pricing the IoT services to maximize the profit of IoT service providers under monopoly and oligopoly IoT service markets. Under monopoly market, the only one IoT service provider can set price to get all the surplus from the users. However, under oligopoly market, there are competitions among the IoT service providers. We model the interaction among IoT service providers as a Bertrand game, Nash equilibriums are established for oligopoly case.

Appendix A

Proof of Proposition 2

We denote the price of n IoT service providers as a vector $\boldsymbol{p} = (p_1, \ldots, p_n)'$, and the price vector $\boldsymbol{p}_{-i} = (p_1, \ldots, p_{i-1}, p_{i+1}, p_n)'$. Since users' QoS valuation is distributed uniformly, we have $F(\lambda_k) = \lambda_k$.
(1) The profit of IoT service provider $\mathbb{I}_1$ can be expressed as follows (The number of users using IoT service of $\mathbb{I}_1$'s can be gotten by **Lemma 2**):

$$\begin{aligned}\Pi_1 &= x_1 \times (p_1 - c_1)\\ &= [1 - F(\tau_k^1)] \times (p_1 - c_1)\\ &= [1 - \tau_k^1] \times (p_1 - c_1)\\ &= \left[1 - \frac{p_1 - p_2}{\phi_1 - \phi_2}\right] \times (p_1 - c_1)\end{aligned} \tag{A.1}$$

To maximize Π_1, we have the following optimal condition:

$$\frac{\mathrm{d}\Pi_1}{\mathrm{d}p_1} = 0 \tag{A.2}$$

Therefore, we can get the optimal price by solving (A.2),

$$\begin{aligned}p_1 &= BR_1(\boldsymbol{p}_{-i})\\ &= \frac{1}{2}p_2 + \frac{1}{2}(\phi_1 - \phi_2 + c_1)\end{aligned} \tag{A.3}$$

The optimal price of IoT service provider $\mathbb{I}_1$ is a function of $\boldsymbol{p}_{-i}$, which is defined as function $BR_1(\boldsymbol{p}_{-i})$. In game theory [7], we call the function $BR_1(\boldsymbol{p}_{-i})$ as ***best response function*** of player IoT service provider $\mathbb{I}_1$.
(2) The profit of IoT service provider, for $i = 2, \ldots, n-1$, can be expressed as

$$\begin{aligned}\Pi_i &= (p_i - c_i)x_i\\ &= (p_i - c_i)[F(\tau_k^{i-1}) - F(\tau_k^i)]\\ &= (p_i - c_i)\big[(\tau_k^{i-1} - \tau_k^i)\big]\\ &= (p_i - c_i)\left(\frac{p_{i-1} - p_i}{\phi_{i-1} - \phi_i} - \frac{p_i - p_{i+1}}{\phi_i - \phi_{i+1}}\right)\end{aligned} \tag{A.4}$$

To maximize Π_i by letting $\frac{\mathrm{d}\Pi_i}{\mathrm{d}p_i} = 0$, we can get the ***best response function*** of player IoT service provider $\mathbb{I}_i$ as shown in (A.5).

$$\begin{aligned}p_i &= BR_i(\boldsymbol{p}_{-i})\\ &= \frac{\phi_i - \phi_{i+1}}{2(\phi_{i-1} - \phi_{i+1})}p_{i-1} + \frac{\phi_{i-1} - \phi_i}{2(\phi_{i-1} - \phi_{i+1})}p_{i+1} + \frac{1}{2}c_i,\\ &\quad \text{for } i = 2, \ldots, n-1\end{aligned} \tag{A.5}$$

(3) The revenue of IoT service provider $\mathbb{I}_n$ can be expressed as follows:

$$\begin{aligned}\Pi_n &= (p_n - c_n)x_n \\ &= (p_n - c_n)[F(\tau_k^{n-1}) - F(\tau_k^n)] \\ &= (p_n - c_n)\big[(\tau_k^{n-1} - \tau_k^n)\big] \\ &= (p_n - c_n)\left(\frac{p_{n-1} - p_n}{\phi_{n-1} - \phi_n} - \frac{p_n}{\phi_n}\right)\end{aligned} \tag{A.6}$$

To maximize Π_n by letting $\frac{\mathrm{d}\Pi_n}{\mathrm{d}p_n} = 0$, we can get the ***best response function*** of player IoT service provider $\mathbb{I}_n$ as that in (A.7).

$$\begin{aligned}p_n &= BR_i(\boldsymbol{p}_{-i}) \\ &= \frac{\phi_n}{2\phi_{n-1}}p_{n-1} + \frac{1}{2}c_n\end{aligned} \tag{A.7}$$

If we define

$$\alpha_{i-1} = \begin{cases} \frac{1}{2}(\phi_1 - \phi_2 + c_1) & \text{if} \quad i = 1 \\ \frac{\phi_i - \phi_{i+1}}{2(\phi_{i-1} - \phi_{i+1})} & \text{else if} \quad i = 2, \ldots, n-1 \\ \frac{\phi_n}{2\phi_{n-1}} & \text{else} \quad i = n \end{cases} \tag{A.8}$$

$$\beta_{i+1} = \frac{\phi_{i-1} - \phi_i}{2(\phi_{i-1} - \phi_{i+1})} \quad \text{for } i = 2, \ldots, n-1 \tag{A.9}$$

IoT service providers' best response functions (see (A.3), (A.5), (A.7)) can be expressed as follows:

$$p_i = BR_i(\boldsymbol{p}_{-i}) = \begin{cases} \frac{1}{2}p_2 + \alpha_0 & \text{if} \quad i = 1 \\ \alpha_{i-1}p_{i-1} + \beta_{i+1}p_{i+1} + \frac{1}{2}c_i & \text{else if} \quad i = 2, \ldots, n-1 \\ \alpha_{n-1}p_{n-1} + \frac{1}{2}c_n & \text{else} \quad i = n \end{cases} \tag{A.10}$$

It is obvious the solution of linear equations in (A.10) is the Nash Equilibrium of the game Γ.

If we can prove that there is a unique solution for linear equations in (A.10), we can conclude that there is a unique Nash Equilibrium of the game Γ.

We can express the linear equations in (A.10) as follows:

$$
\boldsymbol{p} = \begin{pmatrix} p_1 \\ p_2 \\ p_3 \\ p_4 \\ \vdots \\ p_{n-1} \\ p_n \end{pmatrix} = \begin{pmatrix} \frac{1}{2}p_2 + \alpha_0 \\ \alpha_1 p_1 + \beta_3 p_3 + \frac{1}{2}c_2 \\ \alpha_2 p_2 + \beta_4 p_4 + \frac{1}{2}c_3 \\ \alpha_3 p_3 + \beta_5 p_5 + \frac{1}{2}c_4 \\ \vdots \\ \alpha_{n-2}p_{n-2} + \beta_n p_n + \frac{1}{2}c_{n-1} \\ \alpha_{n-1}p_{n-1} + \frac{1}{2}c_n \end{pmatrix}
$$

$$
= \begin{pmatrix} 0 & \frac{1}{2} & 0 & 0 & 0 & \dots & 0 \\ \alpha_1 & 0 & \beta_3 & 0 & 0 & \dots & 0 \\ 0 & \alpha_2 & 0 & \beta_4 & 0 & \dots & 0 \\ 0 & 0 & \alpha_3 & 0 & \beta_5 & \dots & 0 \\ \vdots & \vdots & \vdots & \ddots & \ddots & \ddots & \vdots \\ 0 & 0 & 0 & 0 & \alpha_{n-2} & 0 & \beta_n \\ 0 & 0 & 0 & 0 & 0 & \alpha_{n-1} & 0 \end{pmatrix} \begin{pmatrix} p_1 \\ p_2 \\ p_3 \\ p_4 \\ \vdots \\ p_{n-1} \\ p_n \end{pmatrix} + \begin{pmatrix} \alpha_0 \\ \frac{1}{2}c_2 \\ \frac{1}{2}c_3 \\ \frac{1}{2}c_4 \\ \vdots \\ \frac{1}{2}c_{n-1} \\ \frac{1}{2}c_n \end{pmatrix}_{n\times 1}
$$

If we define matrix $\boldsymbol{A}$ and vector μ as follows:

$$
\boldsymbol{A} = \begin{pmatrix} 0 & \frac{1}{2} & 0 & 0 & 0 & \dots & 0 \\ \alpha_1 & 0 & \beta_3 & 0 & 0 & \dots & 0 \\ 0 & \alpha_2 & 0 & \beta_4 & 0 & \dots & 0 \\ 0 & 0 & \alpha_3 & 0 & \beta_5 & \dots & 0 \\ \vdots & \vdots & \vdots & \ddots & \ddots & \ddots & \vdots \\ 0 & 0 & 0 & 0 & \alpha_{n-2} & 0 & \beta_n \\ 0 & 0 & 0 & 0 & 0 & \alpha_{n-1} & 0 \end{pmatrix}
$$

$$
\mu = \left(\alpha_0, \frac{1}{2}c_2, \dots, \frac{1}{2}c_n\right)' \tag{A.11}
$$

then we can express the best response functions in (A.10) of n IoT service providers as follows:

$$
\boldsymbol{p} = \boldsymbol{A}\boldsymbol{p} + \mu \tag{A.12}
$$

Equation (A.12) can be written as follows:

$$
(\boldsymbol{I} - \boldsymbol{A})\boldsymbol{p} = \mu \tag{A.13}
$$

where $\boldsymbol{I}$ is an $n \times n$ unit matrix with ones on the main diagonal and zeros elsewhere.

By Cramer's rule [9], the necessary and sufficient condition for unique solution of (A.13) is shown as that in (A.14).

$$|\boldsymbol{I}-\boldsymbol{A}| \neq 0 \tag{A.14}$$

And the Nash Equilibrium price is

$$p_i^* = \frac{|(\boldsymbol{I}-\boldsymbol{A})_i|}{|(\boldsymbol{I}-\boldsymbol{A})|} \tag{A.15}$$

where $(\boldsymbol{I}-\boldsymbol{A})_i$ is the matrix formed by replacing the ith column of $(\boldsymbol{I}-\boldsymbol{A})$ by the column vector μ. The operator $|\cdot|$ on a matrix denotes the determinant [9] of the matrix. Thus, $|(\boldsymbol{I}-\boldsymbol{A})|$ is determinant of the matrix $(\boldsymbol{I}-\boldsymbol{A})$, $|(\boldsymbol{I}-\boldsymbol{A})_i|$ is determinant of the matrix $(\boldsymbol{I}-\boldsymbol{A})_i$.

Note that the solution from Cramer's rule is not necessarily positive, therefore, the none positive solution should be omitted.

References

[1] L. Atzori, A. Iera, and G. Morabito, "The internet of things: A survey," Computer Networks, vol. 54, no. 15, pp. 2787–2805, 2010.

[2] D. Munjin and J.H. Morin Proc. IEEE International Conference on Green Computing and Communications (GreenCom 2012), BesanÃğon, France, pp. 156–162, November 2012.

[3] A.E. Al-Fagih, F.M. Al-Turjman, W.M. Alsalih, and H.S. Hassanein, "A priced public sensing framework for heterogeneous IoT architectures," IEEE Transactions on Emerging Topics in Computing, vol. 1, no. 1, pp. 133–147, 2013.

[4] Wikipedia, "Cost-benefit analysis," accessed November, 2015.

[5] D. Uckelmann, "Performance Measurement and Cost Benefit Analysis for RFID and Internet of Things Implementations in Logistics," Quantifying the Value of RFID and the EPCglobal Architecture Framework in Logistics, Springer, Berlin Heidelberg, Germany, ISBN 978-3-642-27990-4, pp. 71–100, February 2012.

[6] J. Tirole, "The theory of industrial organization," MIT Press, Cambridge, Massachusetts, USA, 1988.

[7] D. Fudenberg and J. Tirole, "Game theory," MIT Press, Cambridge, Massachusetts, USA, 1991.

[8] D.M. Kreps and J.A. Scheinkman, "Quantity precommitment and Bertrand competition yield Cournot outcomes," The Bell Journal of Economics, vol. 14, no. 2, pp. 326–337, 1983.

[9] G. Strang, "Introduction to linear algebra," fourth edition, Wellesley Cambridge Press, Cambridge, England, 2009.

Chapter 9

IoT and big data: application for urban planning and building smart cities

Mazhar Rathore[1]*, Anand Paul*[1] *and Awais Ahmad*[1]

Abstract

The rapid growth in the population density in urban cities demands additional, smart, and fast provision of services and infrastructure. Countries and metropolitan authorities are really interested to provide smart and intelligent environment and real-time facilities to their citizens. Citizens also want to be facilitated by the provision of real-time information regarding anything like traffic, flood, security, pollution, etc. To meet the requirements of both the metropolitan authorities and the citizens, we proposed the use of Internet of Things (IoT)-based smart systems for smart city establishment and the urban planning as well. In this chapter, we propose an IoT-based system that uses the massive volume of data, termed as big data, generated by the smart systems to establish smart city and to do urban planning for the bright future. The data are generated from the smart home sensors, vehicular networks, weather and water sensors, smart parking systems, surveillance objects, etc. A four-tier architecture is proposed which include (1) bottom tier-1: responsible for the management and deployment of IoT sources, data generations, and collections; (2) intermediate tier-1: handles all type of communication between sensors, relays, base stations, the Internet, etc.; (3) intermediate tier-2: levers the data management and processing using Hadoop framework; and (4) top tier: is responsible for application and usage of the data analysis, results generation, and smart decisions. The system implementation consists of various phases including data generation and collecting, aggregating, filtration, classification, preprocessing, computing, and decision making. The proposed system implementation is done in Hadoop ecosystem with spark, voltDB, storm or S4 for real-time processing to generate results in order to establish the smart city. For urban planning or city future development, the offline historical data are used using MapReduce programming in Hadoop environment. IoT-based datasets generated by smart homes, smart parking, weather, pollution, and vehicle datasets are used for analysis and evaluation. The system is evaluated with respect to efficiency in terms of throughput and processing time.

[1]The School of Computer Science and Engineering, Kyungpook National University, South Korea

9.1 Introduction

A large number of entities are being linked to the Internet at an unexpected rate comprehending the awareness of the IoT. One of the CISCO reports in 2008 exposed that the number of objects joined with the Internet exceeded the number of individuals living in the world, whereas in 2020, it will reach the limit of 50 billion, causing the expansion of the digital world [1]. There are many diverse domains in which IoT plays an energetic role and improve the excellence of human life. These domains include healthcare, mechanization, robotics, automation, transportation, safety, emergency response, and many more. It has a major impact on handling man-made and natural disasters where it is tough for the human to make decisions. Thus, the Internet is no longer considered as the network of personal computers and servers. However, it has now involved in millions of smart objects in conjunction with the embedded systems. As a result, IoT is significantly growing in its magnitude and scope, giving a new way of opportunities with a lot of other issues and challenges [2]. The majority of the developed countries put forward the national strategies toward the implementations of IoT by making their system smarter with real-time response. For instance, Japan's broadband access is providing the facility of communication between people, people and things, and things and things [3]. Similarly, South Korea's smart home allows their citizens to access most of the things remotely [4]. Next-generation I-Hub of Singapore [5] aims to grasp the next-generation "U" type network through a protected ubiquitous network [6]. Moreover, IoT is growing to be used in healthcare sector [7]. The listed initiatives laid the foundation of IoT [8].

IoT is leading us toward the concept of smart systems, such as smart homes where various electronic appliances are connected with each other with high-quality two-way interactive multimedia services. People are controlling their home appliances from remote location. In the literature, extensive research efforts have been made on the smart home technology [9]. Similarly, the idea of the smart home is also drawn-out toward the smart community where home domain, community domain, and service domain are joined to offer assistances to the humanity. In such system, where a huge number of objects are communicating with each other over the Internet, generating an enormous volume of data, termed as big data. To advance the smart home technology, the enhanced and better analytical techniques of big data can play a main role in the expansion of Information and Communications Technologies (ICTs). Such kind of big data analysis offers a better understanding and valuable information about the future prospects as well as planning and development; thus providing us the insight knowledge about the big data.

IoT not only improves the fields mentioned earlier, but the data generated from IoT devices and smart systems can be used as an asset for citizen's development. One of the documents reported that 70% of the world's population (more than 6 billion) will live in cities and adjacent areas by 2050 [10]. Having such great size of the population, billions of the electrical, electronics, and digital devices will also interact with each other, as a result this will produce overwhelming of big data. Hence, using and analyzing such big data as an asset based on the user desires and choices, the cities would become more developed and smarter. Thus, by running the variations of enabling technologies and advanced big data analytics, the IoT has been leaving its early stages and is arriving into the era of revolution by transforming the traditional network infrastructure into a fully integrated future, the Internet. The massive volume of information generated by the embedded and general devices will be shared across a miscellaneous platform and applications to enhance the cities smarter and foresee accordingly in terms of its planning and development.

Conventionally, for expansion of urban cities, it is an extremely important factor to realize the demand for facility profiling to enhance the proficiency and to bring the recent advancement in the city administration. Currently, a small number of organizations are working with their platforms in order to achieve live monitoring, planning, and gathering urban process parameters. These activities are catered by applying the data collection, data processing and analysis, and decision making. Usually, data collection and real-time processing is costly and difficult to achieve, while providing real-time information and facilities to the citizens. Therefore, there is a need to bring smart technologies that could efficiently generate and accumulate the data, perform real-time as well as offline analysis efficiently, and predict the future for better planning and development. For achieving this, we need useful data generation from valuable sources, efficient and fast collection, and real-time analysis mechanisms and platforms.

Having the understanding of the practicality and potential of the IoT and the smart systems, in this chapter, we propel the concept of smart systems toward the smart city with the notion of urban planning and development achieved through big data analytics. In this chapter, we identified the smart systems that can be used as a source of big data generation, which can be used as an asset for developing smart cities and performing urban planning. For achieving this, we proposed the complete architecture in order to analyze IoT-based smart systems generated big data. The four-tier architecture has the ability to process vast amount of data generating from various sources of the smart system in the city such as smart homes, air gas monitoring sensors, smart car parking, vehicular network. In addition, the complete system implementation model is given, which guides various municipalities to implement the system in order to achieve smarter city. Moreover, the analysis is performed on various reliable smart systems datasets to make smart city decision using the proposed system. Finally, the system is tested and evaluated with respect to efficiency measures in terms of throughput and processing time. The proposed system is implemented and tested using Hadoop framework to achieve parallel processing with spark to accomplish the real-time effects in the case of immediate smart city decision. Moreover, Hadoop with MapReduce programming paradigm is used for analyzing historic big data for urban planning and future enhancements.

The rest of the chapter is organized as follows. In the next section, the motivation behind starting this work is explained. In section 3, the details of the proposed system are given including the deployment of the smart systems, smart devices and sensors, how we can use smart systems data for smart cities development as well as for urban planning. Section 4 typically discusses the analysis details on the urban smart systems datasets in order to achieve smart cities and to do urban planning. In section 5, the system implementation is described at abstract level to guide the municipalities to implement the system. Later, in section 6, the practical system implementation details and evaluation results are given. Finally, in the last section the conclusion and future work is made.

9.2 Motivation

As mentioned earlier, smart cities becoming keener by making intelligent decisions due to the augment nature of digital technology, in which smart city is equipped with different electronic equipment and smart systems, such as street cameras for surveillance system, sensors for transportation system, smart homes, pollution detector sensors. Although there are also various initiatives taken that use objects to provide different value-added facilities, such as street view, global positioning system (GPA), Earth maps. Furthermore, the enriching nature also rises toward the practice of individual mobile devices, contributing in the said scenario. In such heterogeneous environment in terms of entities features, usage, motivations, security rules, etc., different questions arise from a city environment, which need to be responded [10]. These are:

- How to make existing objects smarter? Alternatively, how to design new object smarter based on the user choice in order to facilitate the citizens?
- How to make smart objects to react accordingly with respect to environment and situation?
- How to minimize the cost of data collection, processing, and decision making that is being generated by some devices?
- How to use IoT generated data as a resource in order to get an insight into it, if the data are collected and going to processing stage in a real-time?

Based on the aforementioned questions, the smart city concept exploits ICT by serving the citizens in everyday life within limited resources. Moreover, various organizations aimed at developing a system that uses an innovative technology by providing the proficient services to their citizens. The majority of such modern technologies comprises of cutting-edge sensing abilities, storage capacity for the exceptional volume of data, and finally, to get an insight into the baggy data. However, the earlier research involved in very few research outcomes in the field of smart city development as well as in the better planning of urban areas. Similarly, a solid system is not yet built with more scalability and efficiency. The big data can be used as an asset to analyze various features of the smart city and then uses the knowledge acquired from the past data for better future of the cities. A similar concept is followed using the IoT paradigm and the big data concepts for urban planning. Thus, we tried to come up with a solution that is applicable in the smart city as well as in the urban areas.

In addition, the motivation behind our aim is to supplement the massive deployment of ICT resources in developing the whole system. For this very reason, we acknowledged that the advancement of recent technology in the field of embedded systems depicts the trends of ICT. Therefore, a system is vital that could bring together all of the recent developments in the field of ICT, due to which an extraordinary growth can be seen in a near future. The design of this system entails all the capabilities of sensing the atmosphere and analyzing the sensed information. Moreover, it can be seen that integrating a large amount of data to perform the efficient analysis is already performed at their best. However, with large scale environment, it is inevitable that the huge portion of data are left disjointed. As a result, such data cannot deliver us better knowledge of the situation so that we may make strategies for the near future. For this reason, urban planning and developing offers a new means to the field of the IoT, in which devices are incorporated by means of their geographical location, and they are analyzed by means of freshly designed system for various services in a city.

9.3 Proposed system for urban planning and smart cities

The basic idea of the smart city is to provide right information at the right place and on the right device to anyone who needs it in order to make the city related decision in a well-organized way. The main purpose of smart city is to facilitate the citizens with more quick and fast ways, such as giving current traffic information, pollution information, and any critical event information. Moreover, smart city also aims at providing lots of other information to the authorities to manage the city in a real-time way, such as managing traffic, controlling, and taking action against any crime or any critical incident at real-time. In addition, urban planning also leaves a major impact on a countries development. In the current technological era, the use of IoT for urban planning not only rapid the planning process but also makes the process more effective and intelligent by historical big data analysis. In this section, we provide the complete overview of the proposed system, which describes how we can use IoT-based smart systems in order to generate and use data for smart city building and urban planning. How the sensors are deployed? How the sensors will generate data? These questions are addressed in this section. In addition, the proposed system architecture including its implementation model is presented that have four tiers or layers and system implementation to show the working of the proposed system.

9.3.1 Smart systems deployment and big data generation

One of the basic issues and perhaps many people have the query in mind that how IoT can be used to establish and form a smart city. Thus, we came up with the right discussion and, of course, the solution to this question. IoT is called as the interconnection of heterogeneous devices that tie together over the Internet. Since we are moving forward towards the digital era, smart homes, smart cities, etc., we start thinking that the objects at home and the surroundings, such as washers, refrigerators, TV, traffic control, security surveillance, should be linked to the Internet for fast and remote accessibilities and facilities. In order to reach the objective of smart city establishment, many sensors are being deployed at different places in order to generate and analyze data for better usage. The ultimate goal is to achieve smart homes, smart parking, environment population, smart vehicular traffic, weather and water systems, and surveillance system. To develop the IoT-based smart city concept and urban planning system, several wireless and wired sensors, surveillance cameras, emergency buttons in streets, and other fixed devices are also suggested to be deployed. The main challenge in this regard is to achieve smart city system and link various IoT systems together to get information. This is done by installing relay nodes, aggregators, various classifiers, and gateways. Moreover, all sensors generate rich amount of data with very high speed, which is termed as big data. Therefore, aggregation, collection, and processing of big data are also one of the key challenges. To address the collection and aggregation issue, one or more aggregation servers should be deployed, which collects and aggregates the data from all smart systems. The data are received with high-speed network; therefore, the aggregation process is powerful enough to aggregate the data and send it for analysis through IoT systems. To process and analyze the big data in an efficient way, the Hadoop ecosystem systems are employed. The complete IoT objects deployment is shown in Figure 9.1, in which we can see that various smart systems and other IoT objects are deployed, like smart home, smart parking, weather and water system, surveillance devices, environmental monitoring sensors, and vehicular traffic information network by establishing a complete IoT platform. All the data from each system are gathered by aggregator and sent to the main system.

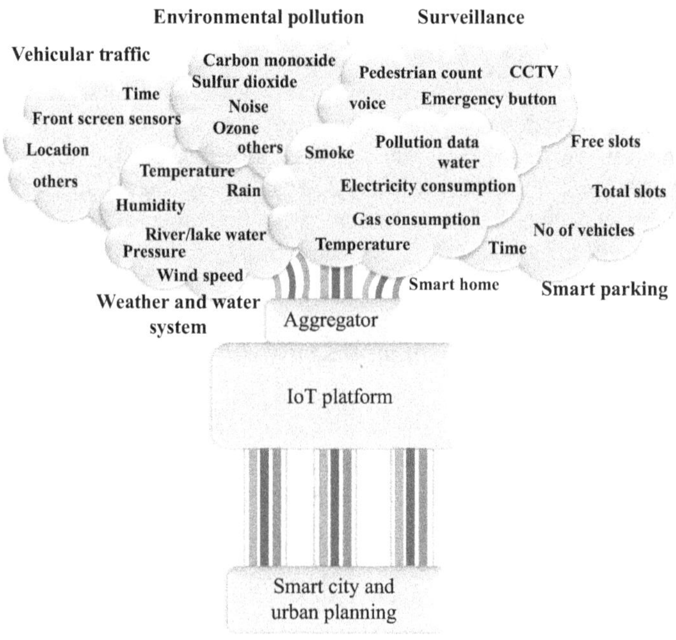

Figure 9.1 Sensors deployment

9.3.2 IoT-based smart city

As we have seen previously, various smart systems and IoT objects are imposed to deploy with the purpose of establishing the smart city. These systems are smart home, smart parking, security surveillance, environmental pollution sensors, weather and water sensor, vehicular traffic network and road sensors, etc. How these systems participate in development of the smart city are described here.

In smart home, the home is continuously monitored by sensors deployed in the house, such as smoke and temperature, water and energy consumption. The smart home generated data can be used in many forms that overall affect the performance of the city, e.g., fire detection at real-time, the electricity and gas consumption to manage the power, gas, and water consumption effectively to the houses as well as different areas of the city. Similarly, monitoring the pollution in the house helps in the healthcare of the households and alerts them when the pollution rises more than a certain threshold.

The smart parking helps in counting and monitoring all the vehicles approaching and leaving various car parking zones. Thus, such types of smart parking system give the information of number of vehicles inside the parking, coming or going out from parking area, etc., which can be used for discovering the places to establish new parking zones. Similarly, the smart car parking data provide lots of other facilitations to the citizens and the merchants also as being a part of the smart city. The nearest free and suitable car parking information can easily be obtained by the citizen in smart city environment. This results in reducing the overall fuel consumption of the vehicles. Moreover, this also saves the time wastage, a person can spend more time in a market place or other activities. In addition, the merchants who are getting less profit are encouraged by diverting the citizens' cars toward their parking area.

Weather and water information also give productivity to the smart city by delivering the weather related information like temperature, pressure, wind speed, rain, humidity, and water levels at rivers, lacks, dams, and other water reservoirs. All these information are accumulated by installing the sensors in water reservoirs for water level or quantity detection and other open places for other weather-related data collection. In all over the world, majority of the floods occur because of heavy rainfall and few others by snow melting and dam breakage. Therefore, the rain monitoring and snow melting sensors are used in order to predict the flood earlier. We can also expect earlier the water quantity in reservoirs in order to meet the water needs of all the citizens. Similarly, we can announce many more things based on critical values of other weather sensors like wind speed, humidity, and others.

Real-time vehicular traffic information is the best source of data for smart traffic management in the city. This data source can be one of the major assets for both government and citizens, if effectively used in order to generate transport-related decisions. The travelers can get to the destination by the quickest route based on the current traffic intensity level and the speed of the vehicles. The traffic can be diverted across various roads within the city, which not only reduces the fuel consumption but also drops the pollution level that occurs due to the crowded traffic. Government authorities, such as traffic police, can also be benefited by locating the blockage of road area due to an accident or other incidents. They can ensure any necessary action based on the current scenario at real-time to manage the traffic. In the proposed smart city system, we are receiving the current traffic information by General Packet Radio Service (GPRS), vehicular sensors, road sensors, as well as the sensors placed at the window screens of the car in order to detect accidents. We can use each vehicle location and the number of vehicles between two pairs of sensors placed at various road junctions of the city. Moreover, in case of any accident, the front screen is normally damaged immediately. In such cases, the sensor instantly sends an alert to the police, traffic authorities, ambulances, and hospitals. Similarly, we can do a lot of other things with real-time traffic data to make the city smarter, more intelligent, and developed.

Pollution monitoring is also vital for human healthcare, especially in this technological era, where there are a lot of transportation, factories, and other sources generating lots of toxic gases. So, delivering the current pollution, i.e. toxic gases level information to the people, is also vital, especially informing heart, liver, asthma patients not to go outside when any gas level is higher. Any metropolitan can never be smarter and developed with unhealthy populations. Therefore, while designing smart city, we place a separate component to get environment data related to toxic gases, such as carbon monoxide, ozone, particular metals, sulfur dioxide, and noise as well. These gases are very hazardous for human fitness, which causes coughing, liver disordering, and heart-related diseases. People with such types of diseases, especially the children, people for physical exercise, already sick people, old age people, should not go outdoors when the level of these gases are more in the air. This can only be possible when people have real-time access to all environmental information and alerts are generated when any of the polluted gas exceeds a certain threshold. Moreover, the places where there are more population, transportation, and factories, government should try to dominate such causes of the pollution like moving industries to other places, diverting traffic to the other routes, etc.

Last but not least, the most significant concern for the people of smart city is the security and safety provision. This is achieved by the proposed system by constantly monitoring the video of all the city area by surveillance cameras. However, it is very difficult to analyze the videos coming from thousands of cameras placed all over the city while detecting any mishap with anyone in a quick manner. To overcome this limitation, we propose a new setup that increases the security system of the complete city. We placed various emergency buttons with microphones at various street areas of the city along with the surveillances cameras. When any misfortune occurs with anybody like robbery, car stolen, rape, fighting, purse stolen, or someone watching some illegal activity in the street, he can just push the emergency button at any nearby place and it will send the message to the closest police station and other authorities. Thus, the police or security agencies start observing the nearby locations through surveillance cameras and can easily trace the imposters. Moreover, the past crime information collected from different sensors can be used to avoid the future security threats by placing more security objects in critical areas. This will lead the authorities to the provision of more secure environment to the residents of the city.

9.3.3 IoT-based urban planning

Same IoT scenario is considered for the urban planning with same devices and sensors, as shown in Figure 9.1. However, the use of sensors generated data and the main purposes of analysis vary in case of urban planning and smart city. In smart city, real-time decision is made based on current data. On the other hand, in urban planning, the historical data generated from the same smart city's IoT devices and systems are used to make planning for future regarding anything related to the city. For example by analyzing the electricity consumption of the previous years, we predict the demand for next year and take necessary action to fulfill the demands.

By smart home generated data, government authorities can analyze previous energy consumption data and growing needs and make future planning for building new dams to produce more energy. Moreover, they can also analyze the pattern of energy usage at different periods and manage the electricity and gas bills accordingly to facilitate the citizens. They can also make energy plans for various periods of the year accordingly. For smart parking and vehicular traffic data, new parking lots needs, new building needs, places to build new roads, or extend roads, all these needs can be planned for the future. The cause of pollution can be analyzed and necessary planning can be made based on the increase or decrease in pollution due to traffic changes. Similarly, analyzing the weather and water consumption datasets, we can make plans for agriculture, for prior flood safety, water reservation, etc. Moreover, based on the temperature data and electricity consumption, we can make a better plan for high-temperature seasons to reduce the consumption of electricity. Similarly, from surveillance datasets, we can analyze the number of crime events, more dangerous place, more effected people, which crime is spreading, etc.; based on these data, the security plans for the next month or even for next year can be set.

9.3.4 Proposed system architecture and implementation model

Based on the needs of smart city and urban planning, we proposed four-tier architecture to analyze IoT big data in order to establish smart cities. The complete architecture is shown in Figure 9.2, where the first tier is the bottom tier, two intermediate tiers, and finally the top tier. Functionality of each tier is described below.

9.3.4.1 Tier-I: bottom tier

This layer is responsible for data generation through various IoT sources and then collecting and aggregating that data. Since there are lots of IoT sensors participating in generation of data. Therefore, a lot of heterogeneous data are produced with varying format, different point of origin and periodicity. Moreover, various data types have security, privacy, and quality requirements. In addition, in sensor data, the metadata are always greater than the actual measure. Therefore, early registration and filtration techniques are applied at this layer, which filters the unnecessary metadata and discards the repeated data.

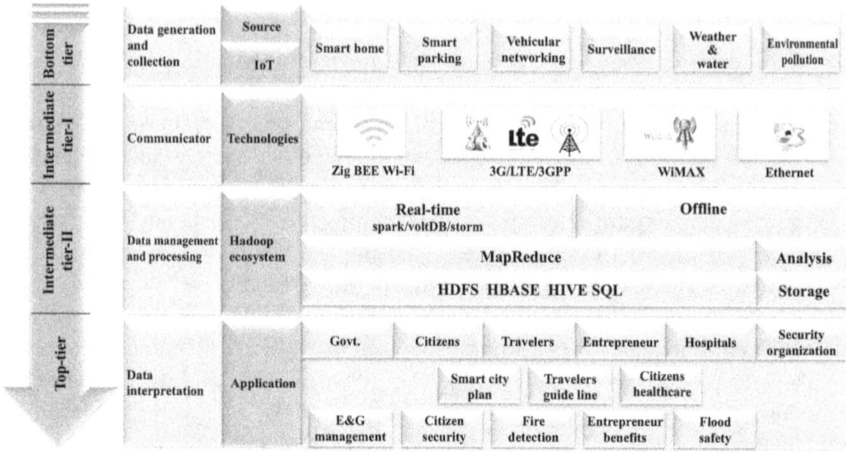

Figure 9.2 Four-tier architecture for IoT big data analytics for remote smart city and urban planning

9.3.4.2 Tier-II: intermediate tier-I

This tier handles the communication between sensors, sensors to relay through ZigBee technology, relays to GW or base station and then on Internet using various communication technologies. At the analysis sides between various analysis servers Ethernet is used.

9.3.4.3 Tier-III: intermediate tier-II

This layer is the main layer of the whole analytical system, which is responsible for the processing of the data. Since we need real-time analysis for smart system. Therefore, we need third-party real-time tool to combine with Hadoop to provide real-time implementation. To provide real-time implementations, Strom, Spark, and VoltDb could also be used. However, for system evaluation, we implemented the system by using Spark. At lower layer of Hadoop, same structure of MapReduce and HDFS is used. With this system, we can also use HIVE, HBASE, and SQL for managing database (in-memory or offline) to store historical information for urban planning, since we care about the efficient results. Therefore, we use Hadoop with the MapReduce programming.

All the data will be stored at Hadoop using HDFS and analysis will be performed at intermediate tier-II. The last tier is the interpretation tier, which uses the results of analyzed data and then generates reports and announcements if required. Here, the generated results are announced and used by many applications, such as flood detection, security, and city planning.

We also designed implementation model of the system, which is shown in Figure 9.3. It shows the complete details of all the steps performed while implementing the system. Initially, every system will generate their own data, such as

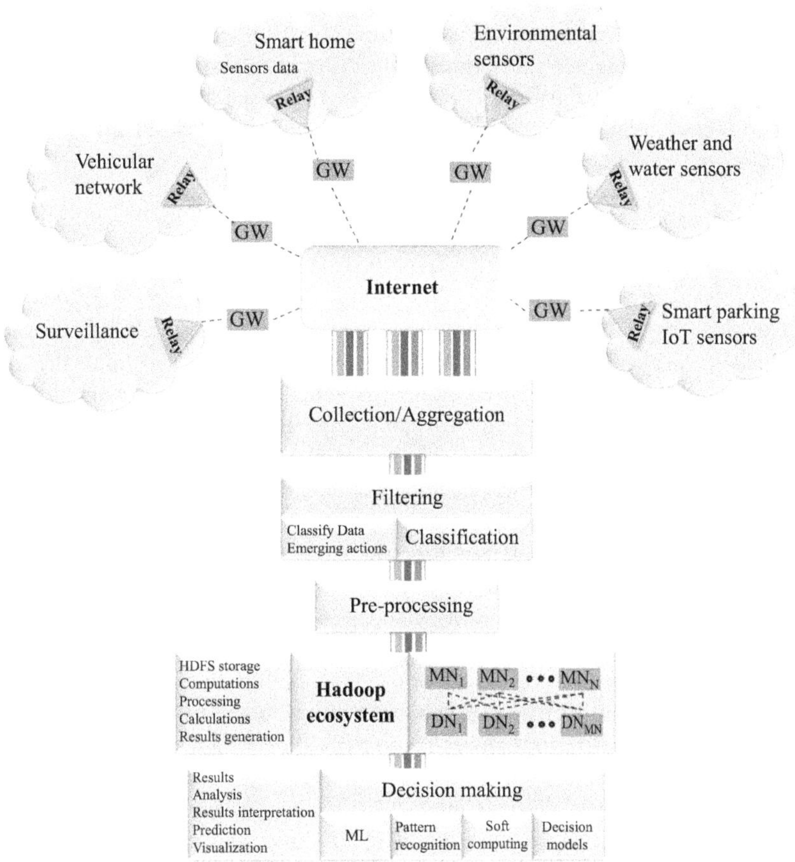

Figure 9.3 Implementation model

smart home generated data, vehicular data, smart parking data. At every system, there is a relay node, which is responsible for collecting data from all the sensors in the system. It uses ZigBee technology to communicate with the sensors. The relay is responsible for collecting data from all sensors, and then sending to the analytical system through GW and Internet. As the sensors have a lot of metadata, the data are of heterogeneous type. Therefore, all the unnecessary metadata and redundant data are discarded. Moreover, the data are classified by the message type and the identifier. After classification, the classified data are converted to the form that is understandable to the Hadoop ecosystem, such as sequence file.

Since we are dealing with a large amount of data (termed as big data), we need a system that could efficiently process a large set of huge datasets. To meet these requirements, we used Hadoop ecosystem, which contains master nodes and various data nodes under the master node. The Hadoop ecosystem has HDFS file storage, which divides the data into an equal amount of chunks and stores them into various data nodes. Later, the parallel processing is performed on these chunks using MapReduce system. All the processing calculations and results generation are done at Hadoop ecosystem. Finally, the decision making is performed based on the results generated by Hadoop ecosystem. The decision-making approach uses machine learning, pattern recognition, soft computing, and decision models.

9.4 Urban data analysis and discussion

To perform the feasibility study and importance of the system, the detailed analysis is performed on various IoT datasets. The analysis is performed to show how the smart city can be built using the proposed system, how the deployment of sensors maters for building smart city, and also how we can use the historic sensors data to perform big data analytics for urban planning. This section also illustrates how we can use the same IoT generated data for both real-time decision making to make a city smarter as well as performing offline analysis on historic data to perform urban planning. In this section, we describe the analysis discussion on various smart systems datasets collected from various reliable sources to establish the smart city and perform useful urban planning for the future of the city.

We take real-world large size IoT generated datasets from various reliable resources. The datasets include (1) the vehicular mobility dataset and road sensors datasets including all the details of the vehicles traveling between many pairs of source and destination points at various places of the city, (2) parking places datasets including the current status of number of vehicles in the parking area, (3) the smart home temperature collected dataset including the water usage of each house, etc., (4) the data of flood occurrence all over the world, (5) pollution datasets including various gases and noise pollution, (6) weather datasets including continuous measurement of temperature, humidity, rain, etc., outside as well as inside the home, and (7) other common city datasets such as cultural events, library events. Few of the datasets analysis discussion is presented here, which shows how helpful the smart system data is, which can be used as an asset for developing smart cities and performing smart planning for the future.

9.4.1 Vehicular traffic analysis

The first dataset that we have analyzed for the smart city and urban planning is Madrid highway vehicular traffic [11]. This dataset is more important for smart city to facilitate the people as well as for urban planning in constructing new roads, building, etc. It contains the location of each vehicle between two edge points of Madrid highway as well as the speed of the vehicle. We also tested the vehicular mobility dataset that is generated by Institute of transportation systems, German Aerospace Center (ITS-DLR) as TAPASCologne project [12–14]. It contains the mobility of all the cars in Cologne city of Germany. It covers the area of 400 km^2 in 24 h with 700 cars. Next all other datasets are covering the Aarhus city of Denmark. In addition, the

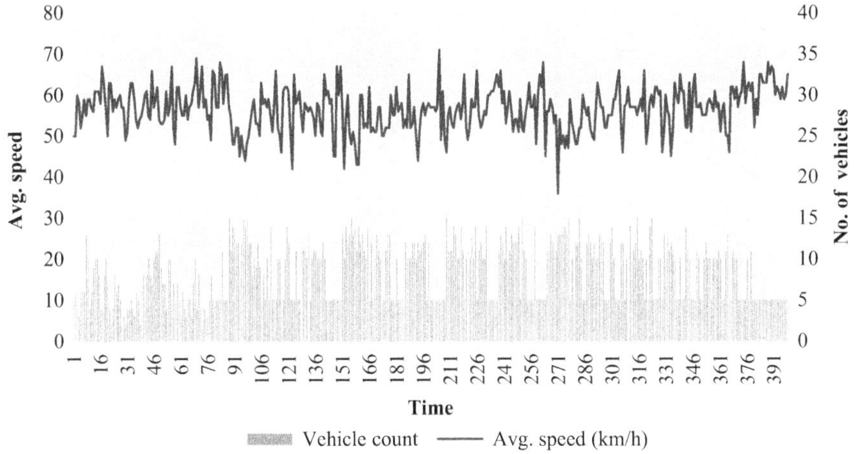

Figure 9.4 Speed of vehicles at low intensity of traffic between two points

Aarhus city vehicular datasets [15–17] are generated by placing sensors at various locations of the city, especially at road junctions. They generate the data by placing 449 sensors at different locations of the city. The sensors are placed at various source and destination points on different locations to estimate the traffic between two points. It contains various information like the average speed of vehicles between two points, the average speed and time to reach the second point. The analysis of Aarhus city traffic is presented only by the data taken from the two sensors placed at 1 km distance in "A rhusvej" street of Hinnerup.

The number of vehicles in a particular area plays a vital role in society. For instance, during the on hours, the traffic intensity at particular roads is higher than off times. Similarly, the road management system can be affected by the number of vehicles in a particular time and on particular roads. In Figures 9.4 and 9.5, we carefully analyzed the traffic intensity on different roads in a society. For instance, if the vehicle speed is low on some roads, then this means that the intensity of the cars is high on those roads. Moreover, in Figure 9.4, when the number of vehicles is higher, e.g. 106 and 121, the vehicle speed is less i.e. 45 and 42. Therefore, keeping this relation between vehicle and vehicle speed, we can design roads for better vehicular management. Similarly, in Figure 9.5, the number of vehicles is taken between 25 and 35, by considering this number as high-intensity traffic. We can see that when the number of vehicles is high e.g. 37, the vehicle speed decreases to 18. Thus, the statistics in Figures 9.4 and 9.5 can be used to design wide roads where the intensity of vehicle is high and vice versa while planning for future.

In Figure 9.6, two types of traffic classes are used, i.e., 1–15 and 25–35 cars. We performed an experiment of reaching moving between two points. We start assuming a car is moving from point A to point B on the road with the number of cars between 1 and 15. The figure shows that the time required for the car to reach its destination is less comparing to the same road with cars between 25 and 35. This estimation is

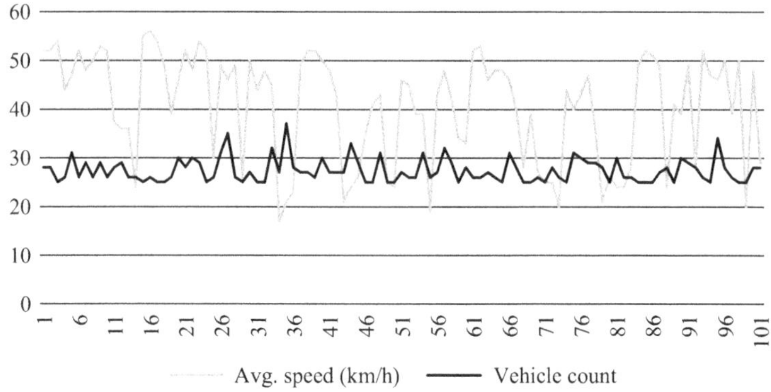

Figure 9.5 Speed of vehicles at high intensity of traffic between two points

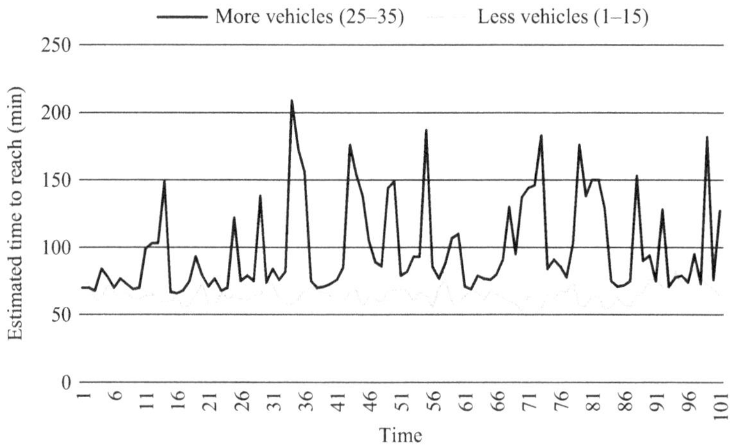

Figure 9.6 Estimated time to reach the destination depending on the traffic intensity

taken at a real-time average speed of the cars running on the roads. Thus, we can design wider roads in those areas where the intensity of car is high. For example, if on a road the number of schools, colleges, universities, etc. is high, then using statistics, a wider road is considered. Similarly, the area where the number of buildings is less than the roads can be designed with less number of lanes. However, we are avoiding the scalability option for now, and we will consider it in our future work.

In Figure 9.7, we check the intensity of the vehicle along a road in a different duration of time. For example, we can see from the graph, during 08:25 and 11:55, the number of vehicles are very high in number i.e. >12. Thus, an efficient road system can design that can dynamically change the routes during the rush hour time.

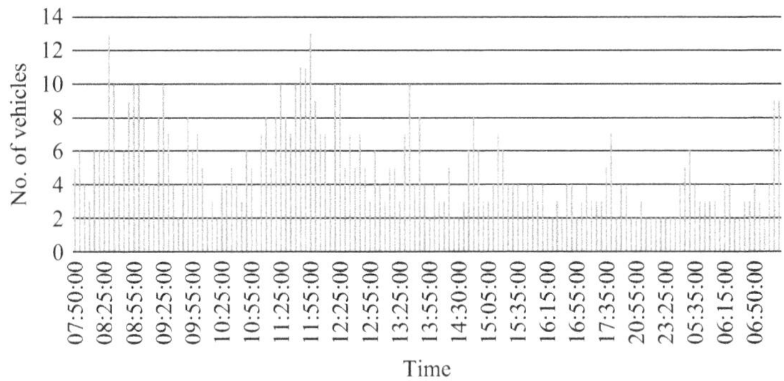

Figure 9.7 Number of vehicles between two source and destination at various time of the day

Similarly, the sensors can be installed at different locations that can communicate with the vehicles in the case of accidents and congestion on the roads. Thus, the various conclusions can be drawn from the statistics of Figure 9.7. For example, the engineer can be provided with better information about the road designing and construction.

From the above IoT-based network traffic analysis, we can predict the estimated time to reach from one point to the other point. Smart city analyzes vehicular traffic data at a real time and facilitates citizens to find how much time it will take them to reach the destination by following alternative routes depending on the current intensity of the traffic. It will give the updated information about all the travelers so that they can make their plan to reach the destination by following the convenient route. Moreover, it also helps the government traffic authorities to control traffic and make optimized plan at run time when the intensity of traffic becomes higher or a road is blocked due to any mishaps that happen on the road like accident, strike, any damage, etc. This traffic management not only helps the citizen and government while providing fuel saving but also provides safety from pollution that is generated by abundant traffic at a single point. So smart city helps the diversion of traffic from busy roads to free roads to get the equal usage of all alternative roads.

In next phase of vehicular traffic analysis, a slightly different dataset covering Madrid is taken. We show the traffic intensity of first 2,500 locations for a particular time in Figure 9.8. The figure shows the congested location where the intensity of the traffic is more. We can easily observe that at starting position there are more vehicles, and when we go forward, the number of vehicles start reducing. It shows that the location 500 is the important location where most of the vehicles are passing through. On the other hand, the location 2,500 is very far from the city, where very fewer cars are moving. Therefore, on the basis of this analysis, we can plan for the road by building more lanes where the traffic is more. Moreover we can also assume that at location 2,500, the number of people living or the number of houses, shops, and building are less; therefore, we can plan to build more houses and buildings there to reduce the traffic burden, pollution, and crown.

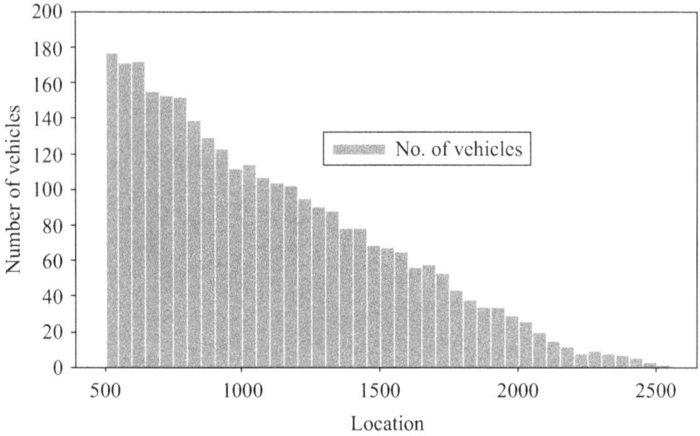

Figure 9.8 Intensity of traffic on various locations of Madrid highway

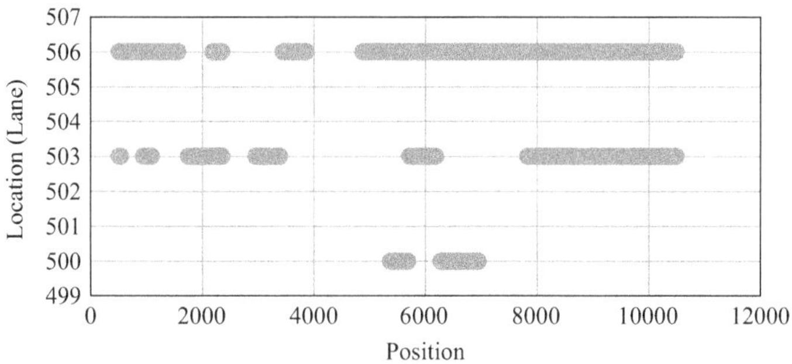

Figure 9.9 Location of speed violation on Madrid highway

For the Madrid traffic data, we also analyzed the speed of the vehicles at the highway. The average speed of the vehicle is 90 km/h. On the basis of speed measured, we can estimate the condition of the road by identifying the regions where the speed of the vehicles goes lower. We can identify the reason for the low speed that cab be the poor structure of the road or the damage of the road. Similarly, in a smart city, we can identify the speed violations of the vehicles at the run time and charged challan on the violation. We identified the regions, where most of the vehicles crosses the maximum limit of speed, as shown in Figure 9.9. On Madrid highway, most of the vehicles crosses the maximum limit from location 5,000 to 1,100. Most of the violations occurs at Lane 3 (most extreme lane) of the highway. These violations might be due to the less number of vehicles on the road. This can be stopped by notifying through sign boards or placing speed barkers at that place, which is suitable for that location. This can also be a better option toward the smart city and urban planning as

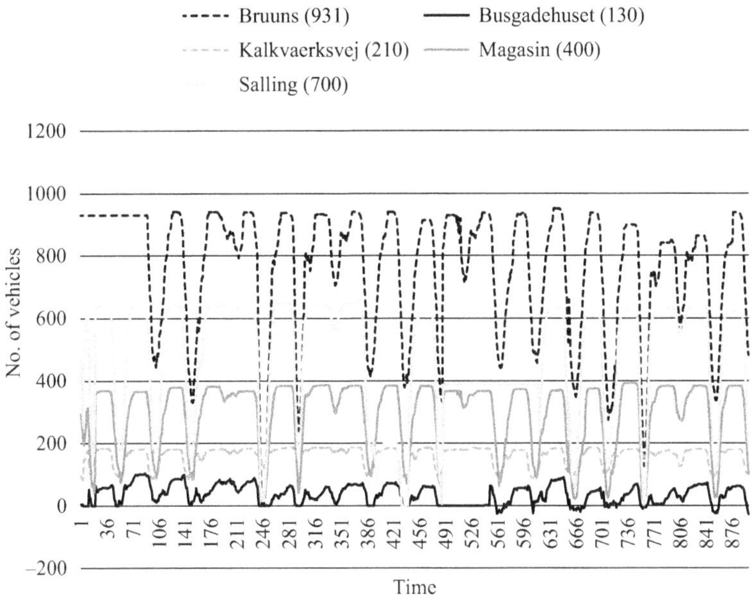

Figure 9.10 Free spaces at various parking lots at different times

well. Moreover, in the smart city, the accident ratio is also monitored with respect to the area speed and violation data.

9.4.2 Smart parking data analysis

Parking lot dataset [15–17] covers the continuous monitoring of eight parking lots of the city with respect to the usage. It contains the data from May 22, 2014, to November 4, 2014, by capturing data through 55 points. By analyzing the parking lots current usage, citizens are updated to select the best suitable parking lot near their location. Figure 9.10 shows the number of free spaces at various parking garages in Aarhus city and Figure 9.11 shows the current use of parking garages. Based on this study, the user can be updated about the free car parking at a run time. He can save his fuel without manually searching the free car garage. Moreover, it also makes profit equilibrium between the sellers in the city by giving benefit to the shop owners who are getting less profit. Normally, citizens prefer to go to the uncongested place for shopping where the number of people is not that much and where they can easily get the parking, resultantly encouraging all sellers. The parking study analysis also gives direction to the government authorities for the urban planning to build more parking areas near the places where most of the people go. In Figure 9.11, it is obvious that the Bruuns is a huge parking area with the capability of parking 931 cars but still you cannot find the parking place few time. This shows the need for more parking lots at that location to facilitate the user. Similarly, the same result can be obtained by the analysis of selling the garage.

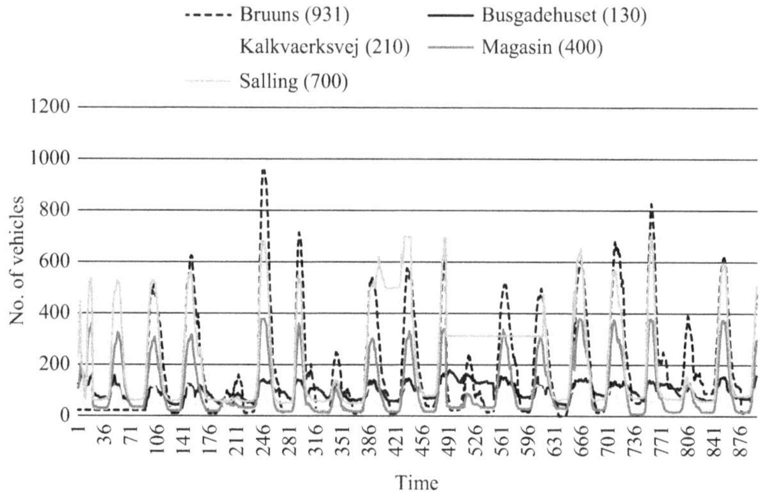

Figure 9.11 Usage of various parking lots at different times

9.4.3 Smart home data analysis

In case of the use of smart home data analysis for smart city and urban planning, we only consider one water usage scenario for illustration. However, it is not only limited to only water usage, there are a lot of other types of smart home data that can be used for other purposes. The water usage data for each household of Surrey city of Canada are taken for household analysis [18]. Water meter readings in a total of 61,263 houses are measured. It contains the complete address and water usage of the house. The water consumption analysis of the household helps smart cities to manage the water resource with respect to the current usage of data. The next year need of water can also be predicted. Moreover, the flow of water to various areas depending on the need of the area can also be controlled. The water consumption of each house of the Surrey city of Canada is analyzed for that purpose. Figure 9.8 shows the histogram of the usage of water in cubic meters at all houses of the city. It shows more than 6,000 houses consume water more than 8,000–9,000 m^3. This shows the normal use of the water at maximum houses.

In general, every city used different amount of water. The consumption of water directly depends on the number of people present in a city. Similarly, some of the cities provide fewer services such as industries, hospitals, universities, schools. Therefore, the population at these cities is less comparatively with other cities. Therefore, using the statistics present in Figure 9.12 helps us in designing the water usage at particular houses within a city. Similarly, the fresh water consumption can be maintained e.g. if a house needs more fresh water and another needs less than a balance relation can be drawn between the houses. Moreover, it also helps the authorities to control the water resources depending on their reservoirs. For instance, if you have more water reservoir, then you can only store the required amount of water by finding the smart

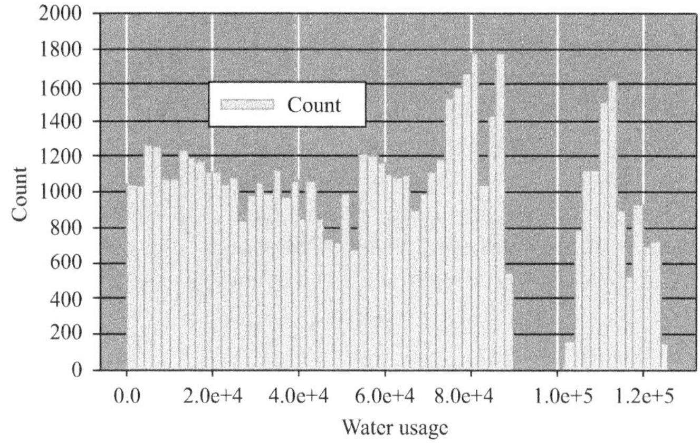

Figure 9.12 Total water usage counts for the Surrey city

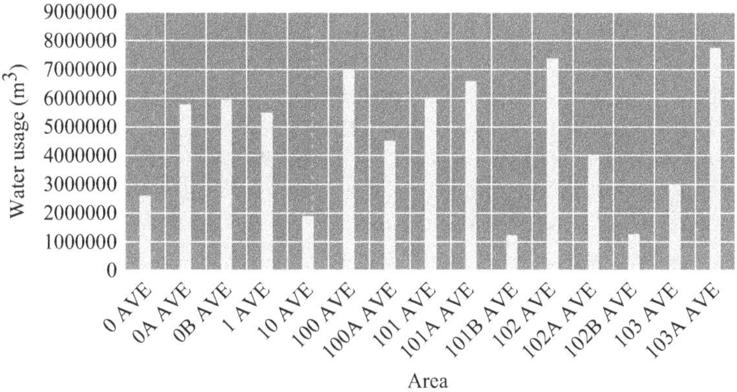

Figure 9.13 Water usage of various areas of Surrey city

city parameters. Likewise, if you have a scarcity of water reservoir, then you can predict the need of water before and then consume the water accordingly.

We also noticed that the water usage in some areas like cities and industrial zones is more than the water usage in the residential area. In Figure 9.13, we show the average water consumption at different areas. For instance, in areas 101B and 102B, the average water consumption is very low. Similarly, in areas 102 and 103A, the average water consumption is very high. This helps us in designing a system by increasing or decreasing the flow and level of water in different areas. Similarly, an efficient drainage system can be designed while keeping the above statistics in mind. Thus, we can draw a conclusion on the basis of water consumption in a particular city by designing a billing system of water usage.

We observed that in a total 61,263 houses, the average consumption of the house is 57877.937. However, 50% of the citizens consume less than 58,186 cubic water (found out by media analysis), 25% citizen use less than 26,893 cubic water. While the 75% of the people use less than 81,983 cubic water. The data are positively skewed means more than 50% use more than average consumption of water. By this analysis of water consumption, smart authorities manage the billing system by choosing a limit for less fixed bill payments and for charging extra amount to those people who consume more amount of water as compared to most of the users. By this management of water, we can manage energy such as electricity and gas as well.

9.4.4 Flood data analysis

For flood detection in smart cities, initially we analyze the reasons behind all the flood in the country. For this purpose, we use G.R. Brakenridge generated the flood dataset [19], which is created by the news from official and TV news channel of the flooded country. The data contain the date of flood, area of flood, damage, intensity, death, etc.

The flood due to rain water normally happens more and intensive as compared to other type of floods such as flood due to snow melting, storm. In Table 9.1, we examine a different type of flood, resulting that the rainwater produces high chances of the flood following by snow. The M represents the magnitude of the flood, which is calculated as, log (duration $\times$ severity $\times$ area affected). For example, if the M value is greater than 4, it means the flood is of a higher intensity. Around 50,250 floods have been experienced with higher intensity at the various areas of the world. Similarly, if the value of M is greater than 6, the intensity of the flood is dangerous. Around 13,751 floods have been recorded of this intensity. The flood ratio in the case of both these magnitude is greater in the event of rain. We can see that 35% of floods have been happening due to the rain following by snow of 1.5%. Thus, we can design a society with predefined thresholds of rain. For instance, if rain in an area crosses a predefined threshold, then a warning signal, or alert can be broadcasted in the society. The society can be made safer by installing high diameter drainage pipe in an area where the rain level is high. Moreover, the rain measure also used to manage the water reservoir in a smart city. Similarly, the snow melting is also a cause of flood but it

Table 9.1 World flood report from 1985 to 2014

Flood type	Total floods	Duration	Total deaths	Total ($M > 4$)	Total ($M > 6$)	Percentage of total floods
Avalanche	3	11	33	14.02157794	0	0.005970149
Rain	3657	41637	190426	17830.89731	6539.589962	35.48437276
Snow	134	2404	851	776.500426	416.4602809	1.54527448
Storm	83	981	6320	473.2605046	229.0867418	0.941811949
Dam break	54	568	3600	163.5712257	44.54054417	0.325514877
Typhoon	5	38	1486	28.63278646	12.34100746	0.05698067

is not that much. This can also be saved by placing snow melting sensors at the hilly station.

9.4.5 Environmental data analysis

Transportation is the main daily activity of the Europeans. Each citizen travel at least 1 h/day [20]. Therefore, a large amount of transportation means such as buses, trains, cars exists in cities. This means of transport cause the emission of 12% CO_2 [21]. Moreover, road population is more than twice as deadly as traffic accidents [22] and car pollution also damaged the youth health and increased the risk of earlier deaths [23]. This shows how much the awareness and safety of pollution are important. The most important gasses in the air that affect the human health are ozone (O_3), carbon monoxide, sulfur dioxide (SO_2), nitrogen oxide, and particulate matter. The environmental existence of these gases is analyzed to deliver the current intensity of those gasses in the air so that more people protect themselves from these gasses.

For analyzing polluted gases in the air, we use pollution datasets [15–17] that are generated by placing sensors at various road points of the Aarhus city. The pollution data have various measures including ozone, nitrogen dioxide, nitrogen oxide, particle matters, carbon dioxide, etc. Finally, the weather data [15–17] consisting of temperature, humidity, rain, pressure, wind, etc., are also considered for analysis and evaluation, which covers the period of February to June and August to September 2014.

Ozone (O_3) is made with three oxygen joined together. It is too dangerous for the living tissues of the human when it comes in contact with them, such as it can harm your lungs, effect to a sunburn inside your lungs, a cough, an irritated throat, or an uncomfortable feeling in your chest, worsened asthma, emphysema and bronchitis, and may reduce the body's ability to fight infections in the respiratory system. It is made with the reaction of volatile organic compounds (VOC), nitrogen oxide (NO), and nitrogen dioxide (NO_2). Therefore, nitrogen dioxide is also dangerous. More VOCs and NO_2 cause more ozone. Sunny weather, less wind, crowded traffic cause increase in ozone. Sulfur dioxide (SO_2) adverse respiratory effects including bronchoconstriction and increased asthma symptoms. "Particulate matter" is a complex fusion of extremely small particles and liquid droplets. The particle can be made by acids (such as nitrates and sulfates), organic chemicals, metals, and soil or dust particles. These are so small that they can get deep into the lungs and cause serious health problems.

For the analysis purpose to keep the gasses value within a limit, the calculations of gas values are a little bit modified [15–17]. However, it will not affect the analysis and reality and effect of the gasses. The values of carbon monoxide, nitrogen dioxide, sulfur dioxide, particulate matter, and ozone index levels gases values are calculated as:

- Initially assigned a value between 25 and 100. Every 5 min, the values will be updated as follows:
  - If the value were below 20 before, it would now be the last value + random integer between 1 and 10.

- If the value were higher than 210, it would now be the last value – random integer between 1 and 10.
- Else the value will be last value + a random integer between −5 and 5.

These gasses are dangerous when their values are greater such as shown in Figure 9.14, the pollution data of Aarhus city are depicted. The maxima values of all gases, as shown as ozone value at time 70–90, particulate matters value at time 185–215 and also at more than 245, nitrogen dioxide at start and end of time interval, and carbon monoxide at 90–115, are all dangerous for health. Therefore, children should not be allowed to spend more time outdoors. Moreover, adults should not exercise outdoor; at that time as healthy persons engaged in physical activity, they breathe faster and more deeply which cause flowing ozone into the lungs. People with respiratory disease should also care when ozone value is higher, as ozone can further irritate the airways of persons who already have diseases of the lung or airways.

For daily based pollution analysis, as we did, we guide the people about the intensity of the pollution and suggested them not to go outside and also do not allow children, diseased persons, and old age people to go out when the intensity of any of the gas is higher. Authorities can also take action and make alert announcements to the public when the pollution goes beyond the limit. The government can also do urban planning by analyzing the history and change behaviors of the pollution in different seasons and months. Overall year analysis, and plan for traffic, city and industrial building can shift industries outside the cities or build new industries at far from cities when these pollution gasses start increasing.

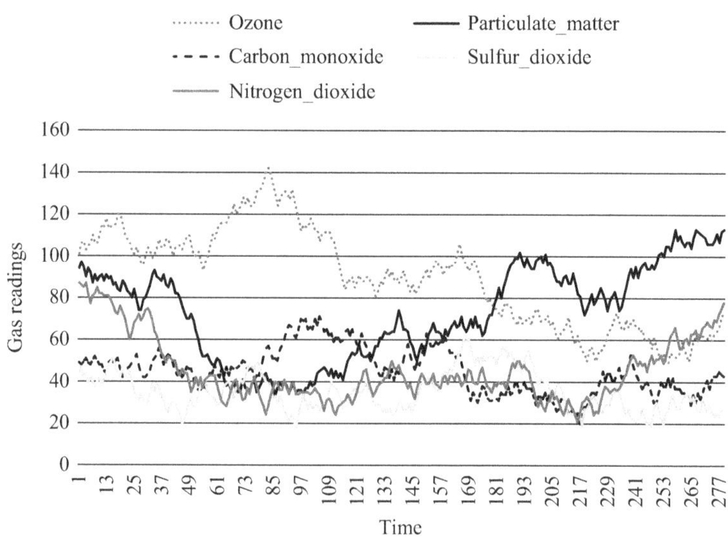

Figure 9.14 Pollution level at different time of the day

9.5 System implementation abstraction

Based on the datasets collected, analysis made, and the proposed system architecture, the system is developed using Hadoop single node at Ubuntu 14.04 LTS with 3.2 GHz × 4 processors and 4 GB memory. The PCaP format traffic is processed by Hadoop-pcap-lib, Hadoop-pcap-scr-de libraries. The network traffic data are then converted into sequence file to make them capable of processing on Hadoop. The system is implemented by two major modules i.e. smart city and urban planning. These two modules further have other submodules for various functionalities. In this section, we provide the implementation at abstract level by describing the modules, source of data, their relationship, and the applications.

9.5.1 Smart city system implementation abstraction

The input source remains the same as described previously as shown in Figure 9.15 with circles outside the boundary of the system i.e. smart home, parking, etc. Each facility of the smart city is implemented as a separated class or submodule, which

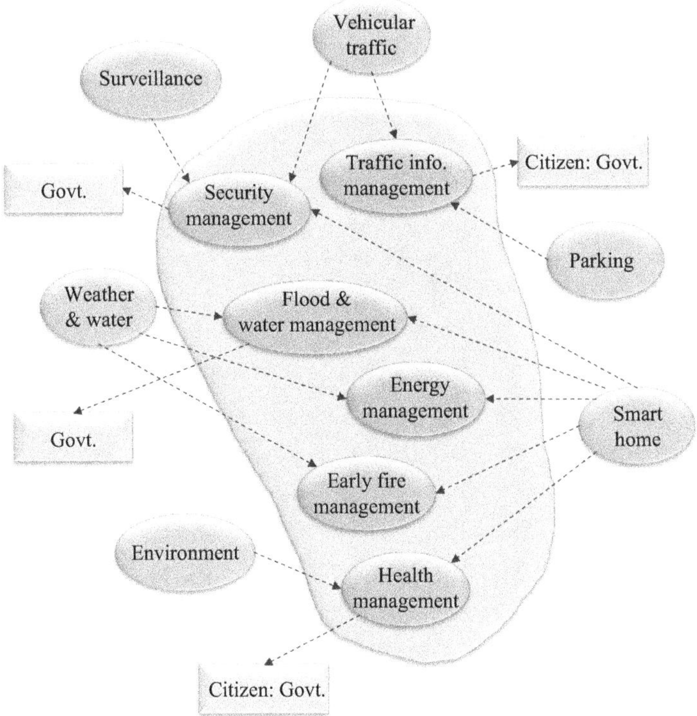

Figure 9.15 Smart city implementation scenario

takes data from various sources. Traffic information measurements take data from the vehicular traffic and parking. Security management module takes data from surveillance, smart home, and vehicular traffic. It takes data from vehicular traffic in case of government needs to monitor stolen vehicles. Flood and water management module take water usage data from smart home rain, storm ice data from weather and predict flood at run time. Similarly, energy consumption management also takes electricity and gas data from smart home and dam and water-related data weather and water. This module will manage and save extra energy that is not used by the several homes. It also distributes energy to various areas according to their needs. Similarly, early fire management program performs fire detection. Finally, the health management makes decision on pollution data. Citizens have limited access to the results of these modules, and the government has full access over them. The complete flow of data, modules, and actors is shown in Figure 9.15.

9.5.2 Urban planning system implementation abstraction

Urban planning system implementation is done in three levels i.e. physical level, intermediate level, and upper level as shown in Figure 9.16. Physical level is called storage level, which is based on Hadoop HTFS system. All the historic data are stored in physical level. Each dataset is given a number in figure such as vehicular data at

Physical level: Storage level	Intermediate level: Processing level	Upper level: Decision level
Historic data	Processing, results	Future planning
1. Vehicular data 2. Energy data 3. Water data 4. Pollution data 5. Weather data 6. Parking slots data 7. Surveillance data 8. Manual annual statistics	1. Statistical measurements (average, correlation, variation, chi-square test, probability calculations) 2. Graphs analysis 3. Other processing	. Road 7 traffic planning (1 : 4 : 6 : 8) . Building, parking, shopping malls planning (1 : 4 : 5: 6 : 7 : 8) . Factories & industries (1 : 2 : 4 : 5 : 8) . Energy need & safety planning (2 : 4 : 5 : 8) . Flood safety planning (3 : 5 : 8) . Environmental health care planning (1 : 4 : 5 : 8) . Security Planning (7 : 8)

Figure 9.16 Urban planning system implementation scenario

number 1, energy data as number 2. Intermediate level is the second level, which is also called processing level. All the processing is done at this level on the data store in the physical level. At this level, statistical calculation, computation, graph analysis, and other computations are performed. The third level is the upper level, which is also called decision level. The decision regarding the urban planning is made at this level. The decision level has various modules for each type of planning, e.g. road planning, building planning. The number written under the planning module is the number of dataset from which the module takes the data for input.

9.6 System real implementation and evaluation

The proposed analysis system is implemented using Hadoop single node setup on UBUNTU 14.04 LTS coreTMi5 machine with 3.2 GHz processor and 4 GB memory. For real-time traffic, we generated Pcap packets by Wireshark libraries and retransmitted them using other systems to evaluate the real-time efficiency of the system. Hadoop-pcap-lib, Hadoop-pcap-serde, and Hadoop Pcap-Input libraries are used for network packets processing and generating Hadoop Readable (sequence file) at collection and aggregation unit so that it can be processed by Spark. MapReduce programming is used for performing offline analysis for urban planning. The dataset mentions in section 4 are used to perform the efficiency evaluation of the system.

Since the system is based on big data analytics, it is evaluated with respect to the efficiency and response time. The system performance is measured various size datasets by considering the processing time (in ms) and throughput (in Mbps). The processing time result is shown in Figure 9.17, and the throughput analysis result is shown in Figure 9.18. It is obvious in the graph that when the data size is increased the processing time proportionally increased, both data size and processing time are

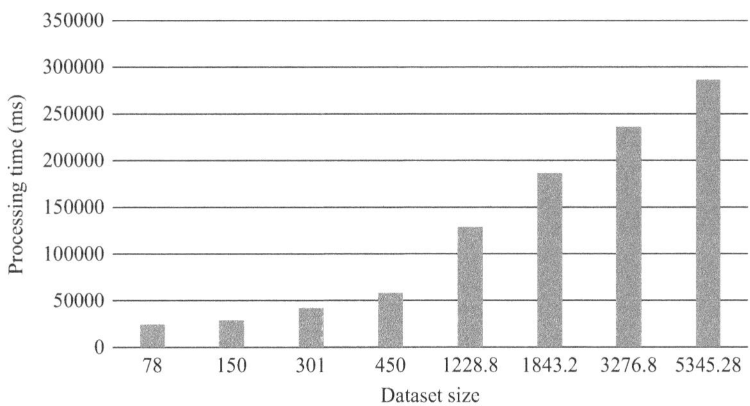

Figure 9.17 Processing time of various size vehicular datasets

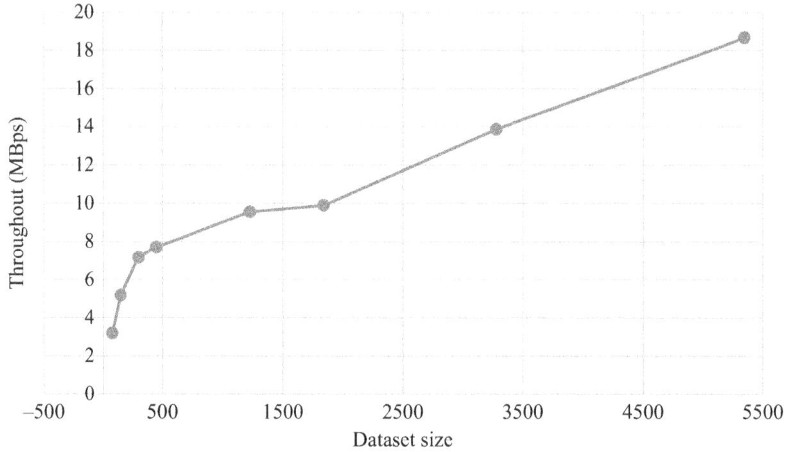

Figure 9.18 Throughput of datasets depending on the size

directly proportional to each other. However, we can examine the processing at higher (larger) dataset i.e. 5,345 MB, the processing time for this dataset is just 300,000 which is far better than other systems. Moreover, when we analyze the throughput corresponding to the data size, we identified throughput is also directly proportional to data size because of the parallel processing nature of Hadoop system. This is the major achievement of the system that with increase in data size the throughput is also increased.

We also check the performance of the system by increasing the number of sensors for a single record. We keep the data size as constant i.e. 2 GB and increase the number of sensors per record, we came to know that with an increase in the number of sensors the throughput is decreased. This is because when we increase the sensors, it will take a lot of time in classification filtrations and processing, as a lot of comparison due to a large number of the sensor in a single record. The throughput of the system with respect to the number of sensors is shown in Figure 9.19.

9.7 Conclusion and future work

Smart cities and urban planning leave a major impact on the development of the nations. It increases the decision power of the societies by making an intelligent effective decision at the appropriate time. It provides convenience to the citizens, so that can get information at real-time without any delay. In this chapter, we propose a system for smart cities and urban planning by using IoT generated big data analysis. A four-tier architecture is proposed which has a collection, aggregation, communication, processing, and interpretation layer. The system describes in detail by designing a complete system implementation model to guide the municipalities to practically

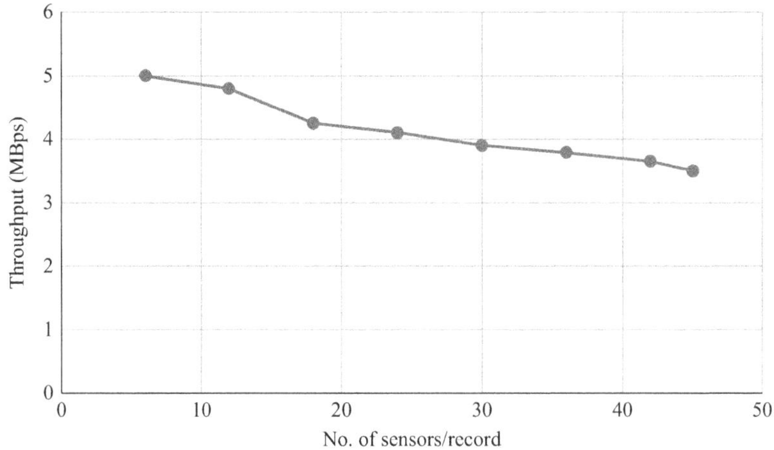

Figure 9.19 Throughput of the system by increase of sensors per record for 1 GB of data

develop the system. Moreover, the system implementation is also given at abstract level by making the system module, the source of data for each module, their relationship, and their application more evident. Finally, the complete system is developed using Hadoop technologies with Spark to achieve real-time processing. The simple IoT-based smart city datasets such as vehicular network, smart parking, smart home, weather, pollution, surveillance are analyzed for making smart city as well as urban planning decisions. The systems are finally tested based on the efficiency performance by considering processing time and throughput. The system gives efficient results even on larger datasets. The system throughput is increased in data size.

In future we are planning to start the practical implementation of the system by developing own smart vehicular transport system. We are planning to use that data for next transport-related planning. Moreover, we are planning to make the city smarter enough to detect and apply smart encounter action when any disaster happens. We want to make the city smarter enough to decide by itself to cater the various problematic situations by making instant and real-time decisions.

References

[1] CISCO, "*The Internet of Things, Infographic*", http://blogs.cisco.com/news/the-internet-of-things-infographic, accessed May 24, 2015.

[2] Z. Deze, S. Guo, and Z. Cheng, "The web of things: a survey". *Journal of Communications* 6, no. 6 (2011): 424–438.

[3] S. Lara, "Japan's ubiquitous mobile information society". *Info* 6, no. 4 (2014): 234–251.

[4] G. Sylvain, and H. Pigot, "From smart homes to smart care" "*ICOST 2005, 3rd International Conference on Smart Homes and Health Telematics*". Vol. 15. IOS Press, 2005.

[5] H. Sun Sheng, "Global city making in Singapore: a real estate perspective". *Progress in Planning* 64, no. 2 (2005): 69–175.

[6] O. Mairtin, and I. Ganchev, "The creation of a ubiquitous consumer wireless world through strategic ITU-T standardization". *IEEE Communications Magazine* 48, no. 10 (2010): 158–165.

[7] A. Ahmad, A. Paul, M. Mazhar Rathore, and H. Chang, "Smart cyber society: integration of capillary devices with high usability based on cyber-physical system". *Elsevier: Future Generation Computer Systems* (in press). doi:10.1016/j.future.2015.08.004

[8] X. Feng, L. T. Yang, L. Wang, and A. Vinel, "Internet of things". *International Journal of Communication Systems* 25, no. 9 (2012): 1101.

[9] D. Sudhir, and R. Prasad, eds. "*Technologies for Home Networking*". John Wiley & Sons, 2007.

[10] J. Jiong, J. Gubbi, S. Marusic, and M. Palaniswami, "An information framework for creating a smart city through Internet of things". *IEEE Internet of Things Journal* 1, no. 2 (2014): 112–121.

[11] M. Gramaglia, O. Trullols-Cruces, D. Naboulsi, M. Fiore, and M. Calderon, "*Vehicular Networks on Two Madrid Highways*". IEEE SECON, Singapore, July 2014.

[12] S. Uppoor, and M. Fiore, "*Large-scale Urban Vehicular Mobility for Networking Research*". IEEE VNC, Amsterdam, The Netherlands, November 2011.

[13] D. Naboulsi, and M. Fiore, "*On the Instantaneous Topology of a Large-Scale Urban Vehicular Network: the Cologne Case*", ACM MobiHoc 2013, Bangalore, India, July 2013.

[14] S. Uppoor, O. Trullols-Cruces, M. Fiore, and J.M. Barcelo-Ordinas, "Generation and analysis of a large-scale urban vehicular mobility dataset". *IEEE Transactions on Mobile Computing* 13, no. 5 (2014).

[15] S. Bischof, A. Karapantelakis, C.-S. Nechifor, A. Sheth, A. Mileo, and P. Barnaghi, "Semantic modeling of smart city data", *Position Paper in W3C Workshop on the Web of Things: Enablers and services for an open Web of Devices*, 25–26 June 2014, Berlin, Germany.

[16] R. Tönjes, P. Barnaghi, M. Ali, *et al.*, "Real time IoT stream processing and large-scale data analytics for smart city applications", poster session, *European Conference on Networks and Communications*, 2014.

[17] S. Kolozali, M. Bermudez-Edo, D. Puschmann, F. Ganz, and P. Barnaghi, "A knowledge-based approach for real-time IoT data stream annotation and processing", *Proc. of the 2014 IEEE International Conference on Internet of Things (iThings 2014)*, Taipei, Taiwan, September 2014.

[18] http://data.surrey.ca/dataset/water-meters, accessed June 30, 2015.

[19] G.R.Brakenridge, "*Global Active Archive of Large Flood Events*". Dartmouth Flood Observatory, University of Colorado, http://floodobservatory.colorado.edu/Archives/index.html, accessed June 30, 2015.

[20] Eurostat, "*Passenger Mobility in Europe*". European Commission, 2007.

[21] Eurostat, "*Energy, Transport and Environment Indicators*". European Commission, 2011.

[22] US environmental Protection Agency (EPA), "*Car Pollution Effects*". 2012.

[23] S. Yim, and S. Barrett, "*Public Health Impacts of Combustion Emissions in the United Kingdom*". Department of Aeronautics and Astronautics, Massachusetts Institute of Technology, Cambridge, United States, 2012.

Chapter 10
Healthcare Internet of Things: fundamental technologies, state-of-the-art standards, and current practices

Alan Diaz[1] *and Wei Wang*[1]

Abstract

Life expectancy has been increasing in the last decades. Although increased life expectancy is one of the main goals of modern society, this involves a high cost for governments, which is distributed among facilities for people with disabilities, health insurance for over 65 years individuals, treatment/medication for chronic diseases, and so on.

The Internet of Things (IoT) proposes a technological approach to assist and alleviate some of those costly problems, and preserve a high quality of life. In this chapter, an updated definition of IoT and the healthcare system is presented. Additionally, a quantifiable representation of current health issues is discussed, in combination with a review of the state of the art of IoT-applications in the healthcare sector that assist these issues, emphasizing in those that concern to energy-aware policies.

10.1 Introduction

One of the main goals of modern society is ameliorating health conditions. There exist diverse diseases that technology has not been able to eradicate. Such is the case of cardiovascular diseases (CVDs), cancers, diabetes, and chronic lung diseases. These are the four main non communicable diseases (NCD), which cannot be transmitted nor are infectious, representing 68% of all deaths in 2012, according to the World Health Organization (WHO) [1]. Only in that year, 17.5 million people died from CVDs, positioning these diseases as the number one cause of death globally. Also, approximately 14 million cases of cancer were detected. Only in the United States, the American Cancer Society, Inc invested $632 million dollars to assist issues including patient support, research, prevention, and detection/treatment expenses [2]. Moreover, in the

[1]Department of Computer Science, San Diego State University, California, USA

next two decades cancer cases are expected to rise to 22 million. Furthermore, there is an estimation of 1 out of 10 adults suffering from diabetes. In 2012, 1.5 million deaths were reported due to a diabetes issue. In 2014, 9% of the adult population was living with this ailment [3]. In addition, chronic lung diseases, such as chronic obstructive pulmonary disease, have taken nearly 3 million people as of 2012, representing a 6% of global death [4].

Similarly, these ailments with high mortality ratio, there are also health disabilities that concern the medical sector and global population. Among the 20 main problems of disability worldwide, there are mental disorders including dementia, epilepsy, and depression. For instance, people over 65 years old are more susceptible to suffer from dementia. This is a population that has been increasing in the past decade, because life expectancy has increased by 6 years approximately since 1990s. The Deloitte Inc Company has estimated that by the year 2050, the population of the elderly is going to be three times bigger. Unfortunately, medical institutions, such as hospitals, do not have the capacity to assist this number of people in their infrastructures. Moreover, the number of caretakers (physicians and nurses) remains static, and studies estimate that this number is going to remain the same in the following years. As we can see, there are still many issues that concern to the healthcare systems that have not been solved, and they are going to get tougher in the coming years.

The healthcare system is in charge of preventing, diagnosing, and treating health issues in an overall population. The healthcare sector is the second largest category as of spending in gross domestic product, representing a 10.5% of waste by 2014, according to the Economist Intelligent Unit [5]. The healthcare systems from all over the world have been benefited in the past decades due to technology. The introduction of computers and medical assistance devices (i.e. Magnetic Resonance Image machines, digital thermometers, etc.) has improved the performance of medical institutions. Patient care and worker efficiency, decision-making assistance with big data analytics, doctors and physicians accessibility, and online consultancy are just a few of the improvements that technology has to offer. In this sense, health assistance can reach a major number of the population, providing good quality service, and improving life-experience among modern society.

Since technology evolved very fast, their application in the medical field has been improving over the past decades. Because of that, the International Organization for Standardization (ISO) released the ISO/TC 215 Health Informatics section, with 146 standards up to date, to support and enable all the aspects of the healthcare system, by developing health information technology (HIT) [6].

Healthcare informatics, which is the area in charge of the development of HIT, is in charge of improving healthcare security, quality, and efficiency by digitalizing all aspects concerned with the healthcare sector, such as electronic health records (EHRs), analysis of data recorded, and better in-time patient assistance.

The IoT is a technology in development that is revolutionizing ubiquitous computing. The proposal suggests that any object can enable a *smart conversation* between another object or group of objects without wires or human intervention. The usage of IoT will enhance scalability and reliability among ubiquitous computing, allowing the development of more efficient and accurate decision-making tools, and user-interacting devices/applications.

Among the features that compose the IoT, there is the capability of constant unobtrusive monitoring. This is a key element for the healthcare system, since it will enhance the quality of the patients' experience. Additionally, enabling objects entering into a smart conversation will inform the stakeholders about the current status of a specific situation, allowing immediate action over an emergency.

In this chapter, a discussion of the most promising technologies that exist to enable the IoT for the healthcare sector is presented, in addition to current applications that have been implemented to assist diverse issues concerning to this area.

This chapter is structured as follows. In section 2, the state of the art of existent technologies that have been proposed to assist the IoT is presented. In section 3, applications of these technologies in real-life scenarios are presented, by introducing previous work of the research community. In section 4, we present open challenges and future directions, and conclusions respectively.

10.2 IoT elements for healthcare

Modern technology has evolved at a significant fast pace. Today, it is common to be in constant interaction with our phones, tablets, or computers, communicating with other devices to *do* whatever we want to, from driving cars to ordering pizza with a simple click on our electronic device. Behind the scenes of this interactive *new* world there is a vast group of different technologies that make this possible. The enhancement of computer networks, the efficiency of nanotechnology and diverse sensors, and re-structuring of computer architecture are allowing the research community to provide society of better tools to assist health conditions. In this section, the state of the existent technologies that have been proposed to assist the IoT issues is presented.

10.2.1 Ambient intelligence (AmI) in general

AmI systems can be seen as a branch of Artificial Intelligence (AI), where the purpose is to generate smart environments to assist common, everyday situations of modern human behavior, in order to improve the quality of life. According to Acampora *et al.* [7], an AmI system must be aware and adaptive to the context it is being implemented in, so it can learn from the environment and be able to anticipate specific needs of an individual. This feature provides a personalized service of AmI systems. In addition, each system must be ubiquitous and transparent, in order to have a big impact without interfering in the user's life.

The difference between the IoT and AmI systems resides in that each application that involves the IoT must follow protocols that include *strictly* the usage of Internet. On the other hand, the diverse applications of AmI systems can use different wireless communication standards such as Bluetooth or ZigBee, just to mention a couple. Further development of AmI includes wireless body area networks (WBANs), wireless mesh sensor networks (WMSNs), and different sensor technologies.

Direct applications of AmI in healthcare include the monitoring of individuals within different scenarios, from the elderly to alert in case of falls, to dementia patients who leave rooms and get themselves in dangerous situations. Similarly, the data analysis and prediction (anticipation) that these systems provide can improve hospital–patient experience among hospitals. Since the information travels between nodes (patients, analysis-systems, doctors) in real-time, a larger number of people can be assisted with better accurate medical results.

Security and infrastructure are the most challenging issues that AmI has to face. In order to learn from the environment, the systems must be recollecting data of the users constantly. The misuse of this information can lead the user/patient into a dangerous situation. Further research is necessary in order to provide this ubiquitous service without compromising relevant-personal data.

10.2.2 Service oriented architecture (SOA)

The implementation of the Internet among enterprises has enabled faster communication between people and electronic devices. This enhancement has been improving over the years, allowing the interconnection of more devices inside large companies. For instance, the development of desktop, mobile, and Web applications in the software industries has improved enormously due to the Internet.

SOA is a software architectural pattern that extends object-oriented architecture. It enables communication between diverse technologies (programming languages, software applications, etc.) through the implementation of services and messages. In this sense, services are software components that can be self-identifiable and application-independent. The messages are encrypted codes that services use to communicate between them. Services communicate through an Enterprise Service Bus (ESB). An ESB is in charge of the routing, reliability, management of services, logging, and security issues of communication [8]. SOA is referred to as a tightly couple technology, which lightens the communication process between applications.

Among the technologies that enable the implementation of SOA, there is the usage of Extensible Markup Language (XML) and JavaScript Object Notation (JSON) that are the most common standards for the development of Web services. Alternatives to these two languages are the Simple Object Access Protocol (SOAP), Hyper-Text Transfer Protocol (HTTP), Representational State Transfer (REST), and Web Service Definition Language (WSDL) among the most popular.

Standard information sharing and scalability are a necessity for today's systems. A SOA infrastructure beneath healthcare systems allows software applications to interact consistently. Consistency is important in the SOA infrastructure, because it provides independency among different programs or processes, which means that individual technologies inside a system could be replaced without restructuring the entire system [9].

10.2.3 Radio frequency identification (RFID)

RFID has been investigated in depth for the last two decades. It consists in a chip/microchip that is composed of *tags* that uniquely identify each device, and a *reader* with an antenna and a transceiver, that allows the reception and transmission of data via radio waves.

There are two types of tags: active tags, that have the necessity of a powered system, a life expectancy of 10 years, and commonly read/write features; and passive tags, that do not require a powered provider, life expectancy is unlimited, are less expensive than active tags, and are widely used for reading purposes. Its frequency range varies between low and high range. More specifically, low frequency band is 100–500 kHz, providing an inexpensive low reading speed; medium frequency band is 10–15 MHz, providing an inexpensive medium reading speed; and finally, high frequency band that is 850–950 MHz, 2.4–5.8 GHz, providing an expensive high reading speed [10].

RFID has been implemented in different sectors such as transportation, technological devices, security systems, the food sector, among others. Since 1996 the ISO has been updating the status of RFID according to modern technology. In the car and food industries, RFID has been implemented entirely. This is because it has helped companies save time and money during industrialization processes. For that reason, many authors consider RFID as the backbone of IoT. Its energy-saving, small size, and bandwidth are the main features that have made researchers and companies elect them as a critical gadget of their services.

The introduction of RFID into the healthcare sector has been challenging due to several issues. Even when it proposes many benefits, for example, real-time data access, time saving, and an improved medical process, it carries interference problems that make them ineffective for a health system, that must be completely reliable. In addition, the implementation of RFID into the medical system has been delayed because of privacy and legal issues [11].

Furthermore, Rosenbaum recollected the main vulnerabilities that RFID has to face in the healthcare area. Cost, Denial of Service (DOS), eavesdropping, possible physical attacks, and spoofing have been reasons that impede the introduction of RFID into the healthcare system [12].

10.2.4 Wireless sensor network (WSN)

WSNs, or sometimes called wireless sensor and actuator networks (WSANs), are telecommunication technologies that allow communication between nodes that are individual and autonomous systems one from the other. There are structured WSNs that have deployed nodes in a preplanned manner; and unstructured WSNs with nodes deployed in an ad hoc manner. WSNs are designed for specific purposes (applications), and are classified in two different categories: for monitoring, allowing the study of specific targets in different areas including the military, business, and health; and for tracking, allowing the reachability or capture of different targets in areas such as military, business, and habitat (animals). WSNs are designed in order to satisfy low energy consumption. For instance, it follows five-layer protocols (application, transport, network, data-link, and physical) to improve communication and the energy usage [13].

WSNs are built in a dynamic network topology. This enhances a scalable and ubiquitous system that is smart enough to cope with node failures due to the independency of each single one, but, this unattended operation can lead to lost data and communication failure. Nonetheless, the heterogeneity that exists among the nodes enables self-correction, which means that in case of a failure, the network will cooperate in order to keep data flowing [14].

These computer networks have been used in several sectors from environmental monitoring and home automation, to healthcare and industrial automation. Due to the fact of its heterogeneity and scalability, WSNs have shown excellent results for the improvement of timesaving, low energy consumption, and commonly low cost implementation [15].

The main challenge for the research community in WSNs is the integration of current technologies to compose scalable and reliable networks. There are different approaches from non-IP-based technologies, such as ZigBee or Z-Wave, to IP-based technologies, such as Bluetooth or Wi-Fi. Up to date, these technologies have been selected to work excellent for different specific areas. Nonetheless, the implementation of WSN for the IoT is still in a development phase.

10.2.5 ZigBee

ZigBee is a counterpart technology from Bluetooth. It enables communication between devices around a Personal Area Network (PAN). It works through a mesh networking topology, and its range varies from the 10 to 20 m (although it can reach 100 m with a significant lack of efficiency). It maximizes battery life and communication efficiency. It was first proposed by the ZigBee Alliance, which is the group that still handles the updates of this technology. ZigBee is composed of the physical and medium access control (MAC) layers, both based on the IEEE 802.15.4 standard, and the network and application layers, defined by the ZigBee specification, which are mostly used in embedded applications [13].

The mesh network topology that rules the ZigBee technology allows strong reliability, because it keeps communicating even when a node cannot enable communication. It can connect several (tens to thousands) devices together. ZigBee 3.0 released in the year 2014 enhanced object's (device's) connectivity. The former improvement has been considered a powerful tool for the development of the IoT. ZigBee presumes having the lowest energy consumption of the products available for IoT, which they called Green Energy. ZigBee 3.0 operates at 2.4 GHz, and uses the networking protocol of ZigBee PRO, enabling device-to-device communication with low-power consumption, low-cost, and low-complexity [16].

Smart home systems and retail systems are the two main targets of ZigBee. After the release of ZigBee 3.0, the ZigBee Alliance has been proposing several applications for IoT in different sectors. The healthcare area is a strong market for this technology, since it enables constant health monitoring and management for chronic diseases, aging independence, general health, wellness, and fitness [17]. For this sector, ZigBee is working in parallel with Continua Health Alliance, which endorsed the ZigBee Health Care as the main technology for their devices' networking. With a vast repertory of applications in different health areas among chronic diseases, independent aging, and general wellness, ZigBee has been gaining reputation for its effectiveness and lower cost, and lower energy consumption, in contrast with current wireless technologies.

10.2.6 Bluetooth

Bluetooth is a wireless standard, first created by Ericsson in 1994. Afterwards, it has been constantly developed and updated by the Bluetooth Special Interest Group (SGI). This technology enables communication between devices, using radio transmission waves at a high speed and with a low cost. In 2002, the IEEE approved the 802.15.1 specification as the Bluetooth wireless technology. Since 2004, Bluetooth's Core Specifications (standards) are handled by the Bluetooth SGI. The former determined the range of Bluetooth, which varies from 1 m (class 3 radios), 10 m (class 2 radios), and 100 m (class 1 radios). Its vast reachability and low power consumption gain investors attention since its first introduction to society [18]. Bluetooth runs over the Internet Protocol (IP), enabling interconnectivity between current known technologies worldwide in a simpler form.

In 2013, Bluetooth 4.1 was released promoting simplest steps to the development of the IoT, providing coexistence between multiple devices, better connections, and improved data transfer. This version has been named Bluetooth Low Energy (BLE) or Bluetooth Smart. BLE, the low-power wireless standard has two device types: single-mode (BLE, Bluetooth Smart) device, and dual-mode (classic or BR/EDR/LE, Bluetooth Smart Ready) device. In addition, BLE best-case scenario could have a maximum data throughput of 5–10 KB/s. This newest release provides device communication in two ways: broadcasting, where a broadcaster sends data constantly, and observers (that can be more than 1) receive this data if they are willing to; and connections, which is the other device-communication that allows the transmission interchange in both ways for two peers only [19].

In 2014, Bluetooth 4.2 was introduced to the family of Bluetooth technologies, enabling higher speed and better security (less interference ratio) [20]. BLE opened a wide area for IoT developers due to its efficiency, allowing software developers and original equipment manufacturers (OEMs) to create better Bluetooth applications and devices. The main advantage of BLE is its energy efficiency, which was designed in order to consume the less energy possible in an exchange-data communication process.

There is an expectancy of three billion Bluetooth devices shipment in 2015. This number is supposed to increase by five billion in 2019 [21]. Fortunately, the healthcare system has been aided by this technology. Currently, there are millions of devices including heart monitors and inhalers that work with this technology. In addition, according to ABI Research there is going to be an annual growth of Bluetooth healthcare devices shipment of 114% between 2012 and 2018. This implies 31 million devices enabled in the healthcare sector by 2018 [22].

10.2.7 IPv6 and IPv6LoWPAN

The IP revolutionized the method of digital data exchange. By identifying each computer with unique addresses and encapsulating messages beneath packages, two devices can have a conversation. IP version 6 (IPv6) and IP version 6 over Low Energy Wireless Personal Network (IPv6LoWPAN) are the newest versions of this protocol. IPv6 provides a 128-bit address space. It supports address 3.4×10^{38} nodes, assuming 100% efficiency. IPv6 provides over 1500 addresses per square foot of the Earth's surface. Hosts are capable of talking to other IPv6 nodes even when some of the infrastructure between them may only support IPv4. Two major mechanisms have been defined to help this transition: dual-stack operation (IPv6 nodes run both IPv6 and IPv4 and use the version field to decide which stack should process and arriving packet) and tunneling (the IPv6 packet is encapsulated within an IPv4 header that has the address of the tunnel endpoint in its header, is transmitted across the IPv4-only piece of network, and then is decapsulated at the endpoint).

The main goal of the IPv6 address allocation plan is to provide aggregation of routing information to reduce the burden on intra domain routers. Its development was intent to support the continued growth of the Internet. From its former official version IPv4 there have been two improvements: auto-configuration and source-directed routing. The IPv4 and IPv6 capabilities have become virtually indistinguishable, so that the main driver for IPv6 remains the need for larger addresses [2].

6LoWPAN defines how to layer IP version 6 (IPv6) over low data rate, low power, small footprint radio networks (LoWPAN) as typified by the IEEE 802.15.4 radio. The name 6LoWPAN was born from the combining of IPv6 and Low Power Wireless Premise Area Networks, but more importantly so that the name would sort to the top of the working groups list. 6LoWPAN is a developing standard from the Internet Engineering Task Force (IETF) 6LoWPAN Working Group and it was designed from the start to be used in small/pico sensor networks [1].

10.3 IoT applications in healthcare

The healthcare sector has been studied in depth due to the fact of unreliable and unaffordable patient care systems' implementation. Several organizations and governments worldwide have been working in conjunction in order to assist the issues that concern specifically to the medical sector. The main future challenges are reducing the costs of production and machinery from medical institutions, providing a high-quality healthcare service for an overall population, assisting the elderly and the population with disabilities or chronic diseases without compromising other human lives, and preserve a green sustainable world through the introduction of *new* technology. In this section, general applications of current technologies addressing an IoT-based solution are presented.

10.3.1 Vital signs

According to ProPublica Journalisms in the Public Interest, in year 2014 the average waiting time at hospitals before seeing a doctor was 24 min in the United States. Because of that, people who may not need immediate medical assistance could be waiting for longer periods of time. Due to this inefficiency, hospitals and other medical institutions are paying high costs to alleviate the patient–hospital experience.

Among the first activities that medical caretakers must complete when a patient first arrive to the medical institution is retrieving the vital signs of the person in question. This must be performed due to a medical record storage system that aids to prevent or assist future irregularities in one's medical record.

The current process of vital-signs-retrieval in the average hospital consists in a group of repetitive and exhaustive steps, with the usage of diverse devices, and commonly with a handwritten recording method. HIT has provided the automation of different medical services, mostly, the digitalization of medical records. Due to this enhancement, the hard work used to be done by medical assistants has been significantly reduced. In addition, the introduction of electronic medical records (EMR) or EHRs, brought benefits such as providing flow reports, emergency-alert systems, graphs/charts, electronic medical consultancy, and description support tools [23]. The US government created the meaningful use incentive program, which finances and regulates the usage of EMRs all over the country.

A primary step for EMR composition is the retrieval of data. Sensors' capabilities provide diverse features to this inquiry. Rolim *et al.* [24] proposed a WSN-like approach that assists the recollection of patient's vital signs through a consistent method. The system implements different sensors *on* existing medical devices to retrieve patient's data and communicate to other hosts or stations to analyze and manipulate the information. The protocol of communication used by Rolim *et al.* is Wireless 802.11, exchanging XML and HTTP format files. At the same time, the system provides cloud-computing capabilities for the storage and analysis processes. In this sense, patient's information can be accessed in real time from anywhere, providing medical assistants to improve their services.

Similar to the EMR system proposed by Rolim *et al.* [23], Méndez and Ren designed a basic medical data recording system with limited conditions and less scalability. Nonetheless, this system enhances nurse's data-recollecting activity by retrieving patient's data from ECG sensors, oximeters, thermometers, and pulse meters. Méndez and Ren decided to utilize commercial existing technologies such as Spot Vital Signs LXi® for the sensors, an EMR system known as Centricity Practice Solution® developed by GE Healthcare® for the interface, and Citrix Servers which comprises all the cloud storage functionalities. In contrast to other existing cyber-physical systems for vital signs recollection, this enhancement uniquely recollects and stores patient's data, it never analyzes it. Consequently, it reduces the time an individual has to spend at a hospital, but still has further improvements to come. It communicates the sensors with the EMR system through the Internet, but not wirelessly.

In addition to vital signs recording features, the emergencies-area of medical institutions has had many improvements due to HIT. Specifically, ambulance–hospital communication through Internet-wirelessly-technologies could aid to the reduction of physician's procedural workload, enabling them to improve performance in critical situations. There are a few commercial systems that have been implemented in real hospitals. But further research is required due to costs reduction and better communication performance.

Although EMRs have improved the patients' record management among institutions, Personal Health Records (PHRs) propose optimized information management and security services due to lighter and simplified capabilities. In PHR systems each user/patient has total control over personal data access/sharing [25]. Beyond PHR, a tendency of Emergency Medical Systems (EMS) was raised, which in contrast to EMR or PHR systems, EMS operates with ubiquitous real-time capabilities, addressing personal data security issues with cloud-computing protocols. Koufi *et al.* [25] discuss the implementation of EMS through a PHR interface, and cloud computing access management. The flexible, scalable infrastructure proposed all-time-anywhere accessibility to health records. Scalable in a sense that it can operate independently of diverse platforms, and at the same time provide security to sensitive data. Their system *NefeliPortal* is a prototype portal application that enables access to a PHR-cloud via web services deployed through Business Process Execution Languages (BPEL).

SOA systems that guarantee coverage of all security issues are highly irregular. Similar to the NefeliPortal application, Poulymenopoulou *et al.* [26] proposed an electronic emergency personal record (E-EPR) system in charge of managing patient's information in critical scenarios. Its implementation was intended for the use of ambulance–hospital systems, E-EPR ubiquitous functionalities enhance its scalability and platform independence due to its security components. In order to satisfy the Health Insurance Portability and Accountability Act of 1996 (HIPAA) privacy norms, and still provide a high-quality service, this system's security steps are mount on the Amazon cloud infrastructure. In addition, instead of performing all the data storage and analysis through a single cloud structure, the authors proposed two different servers: one for the patient or ambulance station, and the other for the static station (hospital, laboratory, etc.). This E-EPR architecture organized domain context information into domain ontology. The former approach enabled context sharing in a semantic manner with context reasoning. By addressing server-error-prone situations, this system guarantees communication's reliability.

There are several cases of domain-oriented service computing, origins of cloud computing, which manipulate medical data on server side. Among the information that is retrieved from patients there are image files. Images that are usually in the Digital Imaging and Communications in Medicine (DICOM) format, which provides vast data about specific inquiries (personal records, X-ray results, mammography results, etc.). For this purpose, several processing and analysis is required from the doctor and physicians in order to provide an accurate response to the patient's problems. Performing the analysis from the server side is exhaustive and high costly when the number of images increases. Huang *et al.* [27] proposed a medical information integration based cloud computing center that through the usage of web-services provides medical image data storage and analysis at an application level. This system utilizes the Apache Hadoop open-source software framework for the image storage features and analysis. Adding the MapReduce framework capabilities, this system provides accessibility to relevant data from anywhere in the world. The key feature of this medical information center is the development of all the heavy processes on the server side.

Generalizing in healthcare disease-assistance systems development is quite hard due to irregularities that are present in most chronic illnesses. The level of intelligence of current systems, even when high, does not provide a percentage of reliability that allows technicians to introduce them into a full time healthcare monitoring system. Nonetheless, several enhancements have been proposed to assist issues such as blood glucose and digestive system monitoring.

For assisting diabetes mellitus record monitoring, Jara *et al.* [28] designed an ambient assisting living (AAL) application based on the IoT, which alleviates patient's constant monitoring by connecting the glucometers through IPv6LoWPAN protocol and other technologies such as Bluetooth and RFID tags. Blood glucose levels are recorded into RFID tags that are later used by physicians in order to verify blood glucose levels and insulin therapy of the patients.

The development of swallowable capsules has been around since late 1970s. Major approaches were proposed in the early 2000s. Because of those, a whole different research area was open, addressing mini robotic technologies. The original motivation was to assist the embarrassed and uncomfortable process of endoscopy or colonoscopy studies. Once utilizing RF sensors and high quality cameras addressed this solution, further technology has been developed to assist strong diseases that concerns to the digestive system. In addition, developed work such as the *Vector Project* [29] have been studying the capacities of capsule-size robots that can operate inside the human body. Furthermore, in 2010, Olympus Medical Systems Corp and Siemens Healthcare developed a magnetically guided endoscopic capsule (MGEC) for gastrointestinal endoscopy, which operates with wireless technologies [30]. Also, Caprara *et al.* [31] designed a capsule with water propulsion capability, which enhances capsule-camera visibility, in order to assist gastric cancer. In this case, the camera communicates to a desktop station wirelessly.

10.3.2 Smart drug intake

The healthcare sector has to deal with capacitances of drug intake in order to prevent cases of adverse drug reaction (ADR), which its prevalence is of 6.7% throughout hospitals worldwide. Even when the rates of mortality due to ADR maintain low, there is a necessity to eradicate this problem. Not deadly situations of ADR tend to lead to lifelong issues that are uncomfortable and expensive. In addition, wrong drug intake could lead to chaotic scenarios and deadly circumstances.

A typical scenario resides in over taking or not taking the necessary dosage of drugs as prescribed by doctors or physicians. Cases of incomplete treatment are extremely common once a person self-evaluates her/his relief. This misbehavior leads to unsuccessful recovery, and in chronic cases treatments, to null improvements. On the other side, there is the over dosage, which in the long run reduces general drugs' effectiveness.

Only in the United States, an amount of US$177 billion could be saved from the medication noncompliance area. Due to this fact, there is a need to implement smart systems that assist this issue. For instance, Jara *et al.* [32] in collaboration with the WHO have developed a multitechnologies system to improve drug delivery service, providing a guarantee that a high percentage of the population has access to medicines. Also, the system enhances appropriate consumption by tracking drug–patient activity. The system strongly avoids clinical caused by dosage mistakes. The system, *Movital*, establishes functionality through RFID MiF are Desfire tag systems, which communicate through the IPv6LoWPAN protocol, improving effectiveness and reliability of the pharmaceutical sector. In general, Movital provides an identification system for medicines that provides further information of the dosage and usability of them. Although RFID achieves excellent performance, its implementation is affordable for the high-income countries of the world. Due to costs concerns, Jara *et al.* additionally proposed the same system's structure with different and less expensive technologies. The Infrared Data Association (IrDA) interest group releases protocols for wireless infrared communications. By taking advantage of IrDA TBTag, the costs diminished significantly in comparison with the RFID. Other technologies not as expensive as RFID MiF are Desfire systems, but less economic than IrDA TBTags are RFID Ultrafire and RFID I-Code SLI.

Similarly, a pervasive and preventive system for medication noncompliance and daily monitoring was proposed by Pang *et al.* [33]. The intent of this approach was to provide a smart *Pervasive Healthcare Station* in home. It is composed of three different parts. First, an *iTag*, which is an unobtrusive wearable device, composed of different sensors, which constantly retrieves vital signs and communicates to the main system. Second, an *iPackage*, which is an electronic container for medicines that controls patient's access to medicines (i.e. denies access in case of tentative overdose). Third, the main system, which is a 2D-Mesh-NoC based multicore architecture that enables data analysis and communication. Controlled Delamination Material (CDM) seals the iPackage and it is controlled by RFID. The iTag communicates with the system using ZigBee technology, while the main system establishes Wi-Fi connection to store and analyze data.

Similar to the Pervasive Healthcare Station, López-Nores *et al.* [34] designed an *iCabiNET* that perform the same drug identification, but also performs reminder activities. Once a drug is registered in the iCabiNET for further consumption, the systems communicates with the user in order to perform a reminder of the intake. iCabiNET, as the Pervasive Healthcare Station, identifies each drug through RFID tags. On the other hand, it utilizes Bluetooth from a mobile phone to establish connection to OSGi-based residential networks.

10.3.3 Elderly care

One out of three adults aging 65 or older has a fall accident each year. These are the leading cause of fatal and nonfatal injuries. The medical costs of these accidents were US$34 billion by 2013 only in the United States. In addition, there are over 258,000 hip fractures each year due to falls, which is a condition that leads to a dependent living, and in some cases, to an early death [35].

The elderly population has been increasing considerably in the past decades. Research stated that by 2030, there would be 72.1 million older persons in the United States, which is almost two times the population of 2012 [36]. This highlighted the necessity of effective fall-detection systems to assist cases immediately, in order to achieve increase life expectancy rates, and significantly reduction of medical costs.

Unfortunately, the older population also suffers from diverse diseases that obliged them to live under constant monitoring. Dementia is among the main disabilities that concern to the elderly. According to *Alzheimer's Disease International* the number of people living with dementia in 2014 was estimated at 44 million worldwide, and it would be almost the double by 2030, and triple by 2050. The global cost of dementia was about US$604 billion in 2010 in the United States [37].

Assisting the former needs, the research community has been involved in several projects. Nonetheless, only a few proposals avoid obtrusive applications. In addition, the development of smart systems to detect falls is still quite a novel field of study. The IoT and the implementation of cyber-physical systems intent to develop fall-detection systems preserving a high quality of life for the patients, and keep the costs as low as new technologies enable.

WSNs have gained the main popularity among all proposals because they manage large scales with smaller devices. For example, a pervasive computing model based on WSN is presented by Díaz-Ramírez *et al.* [38], assisting patients with dementia through monitor features. If a patient leaves a safe space, the system is designed to emit alerts to the caregivers. This WSN uses binary sensors to detect the events, and are combined with passive infrared sensors and magnetometers.

For instance, IoT intend to communicate the physical world through digital/cyber options. Further studies have proposed different changes in current technologies in order to optimize them for cyber-physical systems. Wang *et al.* [39] discuss a secure healthcare application architecture: cyber-physical enhanced secured WSNs integrated cloud computing for u-life care (CPeSC3). By implementing a communication core, a computation core, and a resource scheduling/management core, Wang *et al.* described a system where medical institutions establish communication with patient's homes in order to detect and assist health emergencies. The authors communicating information through the Internet also proposed the usage of RFID tags or ZigBee devices for the interchange of data between monitoring devices.

Among the commercial options there is *iWander*, an application for Android mobile devices proposed by Sposaro *et al.* [40] to assist dementia patients and caregivers. This approach utilizes a Bayesian network to interchange data, and determine whether the behavior of the individual is normal or abnormal. The simplicity of this project relies in the user-friendly unobtrusive application feature of using personal mobile devices (i.e. smart phones, watches, etc.).

Ubiquitous healthcare (u-Healthcare) is an emerging approach for the healthcare system. This technology proposes the development of ubiquitous computing applications that assist the medical sector. The usage of IP6LoWPAN or 3G/4G/Wi-Fi communication protocols enables the development of ubiquitous applications since they enlightened and improve the process of data interchange. Tabish *et al.* [41] discussed a u-Healthcare system that monitors with real-time capabilities. The strict usage of the newest version of IP alleviates the listening time of the server side and enables the great velocity service.

The capabilities of the global positioning system (GPS) have been widely studied by the research community. This feature is mostly found in every mobile device. A mobile device has multiple features that interact with the real world and permit the detection of different information. Detection processes are enabled due to hardware facilities such as accelerometers and weather sensors. Kau and Chen [42] discuss a fall-detection system that is based on smartphones that are placed in the pockets of low-body clothing. Once their algorithm detect a *real* fall, the application sends an alert to the user's caregivers using 3G network and Wi-Fi technologies.

Body-mounted embedded Systems on Chip (SoC) devices are affordable alternatives in patient's monitoring. These could be watches, earrings, necklaces or bracelets with accelerometer capabilities and micro-controller circuits. Rakhecha and Hsu [43] used a smart watch, in which features are programmed on the micro-controller circuit to distinguish between activities of daily life (ADL) and genuine falls. The latter avoids triggering false alarms, and permits the user to perform regular everyday activities. In this system, all the users must wear a personal watch, because communication is establish wirelessly (radio waves) through these devices.

The development of intelligent systems for these scenarios is a hard task due to the number of variables that must be taken into account. Genetic programming (GP), which is a powerful evolutionary learning method, has helped in some projects to add intelligence to the systems. GP proposes the minimization of human expert intervention since it works on raw sensor input, not relying on manually selection features. This implementation has been studied by Dau *et al.* [44]. By utilizing smartphones communication capabilities the cost can be reduced significantly. Further research is required in GP in order to combine its functionalities with other IoT features.

10.3.4 Healthcare applications of AmI

AmI was first proposed by Philip Research back in 1999 [45]. Since then, Philip's alliance with the Massachusetts Institute of Technology (MIT) and the National Institute for Research in Computer Science and Control (INRIA – Institut National de Recherche en Informatique et en Automatique) motivated the development of applications in the area.

This approach has attracted the attention not only of the research community, but also of large companies worldwide. There are several plans such as 2020 or 2030, where it is expected that AmI systems will be running in our everyday environment.

For instance, the AmI term refers as first mentioned by Philips as "people living easily in digital environments in which the electronics are sensitive to people's needs, personalized to their requirements, anticipatory of their behavior and responsive to their presence" [45]. In this sense, an intelligent system must be in constant communication with the environment, managing real-time updates, and perform complex analytics. Although AmI are directly connected to AI systems, they are not synonymous and operate differently [46]. In general, AmI systems are more complex and implement diverse technologies not utilized by AI.

AmI has many research areas: *AmI and Well-being, Social Robots, Evaluation, City*, among the most popular. This section is concentrated in *AmI and Healthcare*. This field investigates and evaluates different technologies that comprise the development and implementation of AmI systems in the healthcare sector.

For this purpose, the following are actions that each AmI system applied in healthcare must perform: activity recognition, behavior discovery, anomaly detection, planning, decision support, and anonymization [47]. Although many technologies enable the development of this type of systems, only a few of them allow optimization and reduction of costs in both development and implementation. Approaches such as Bluetooth BLT and ZigBee 3.0 provide capabilities for these purposes.

Furthermore, according to Acampora *et al.* [47], among the main applications in the healthcare sector are: continuous health monitoring, continuous behavioral monitoring, monitoring for emergency detection, assisted living, therapy and rehabilitation, persuasive well-being, emotional well-being, and smart hospitals. In addition, the principal concerns of these systems are security issues. Because of that, a *privacy-by-design* (PbD) model was proposed by Langheinrich [48], which only enables the sensor systems once all the user's privacy requirements have been analyzed. In addition, social and ethical concerns have been delaying the implementation of AmI in our everyday environment, due to a lack of reliability on the privacy and security features. The latter has gained the attraction of the research community in order to develop new protocols that preserve individual intimacy.

A WSN system designed for long-term health monitoring in assisted living environments was proposed by Wood *et al.* [49]. The system *AlarmNet* was designed with two main features: individual adaptability to the environment providing feedback constantly following the privacy policy, and extensibility, which enhance ubiquity and robustness. Diverse technologies enabled the implementation of AlarmNet, overall, sensors' application for mobility and emplacement provides flexibility. This personalized system restricts access to all the stakeholders due to security/privacy concerns. A strong characteristic of AlarmNet is its development by *modules*. In addition, its implementation considered ZigBee-based sensors for monitoring and communicating each module.

A similar approach to AlarmNet is proposed by Lu and Fu [50]. One of the main differences between the two systems is that the latter communicates its network through ambient-intelligence compliant objects (AICOs). These focus on *estimating* resident's activities. Also, instead of controlling management via different modules, it controls through different *layers*. This system integrates strict concepts of IoT, since it placed an AICO in almost each element (device/object) of a house environment. Through this AICO implementation it is possible to identify both objects and locations. Due to the identification capabilities, it is easier and more efficient the detection of certain activities and irregularities. At the core of this system is the implementation of a Bayesian network with multiple naïve Bayes classifiers, which improves control over error-prone wireless sensors.

A main challenge when developing AmI systems is the different communication protocols among diverse technologies. Although the IP (especially IPv6LoWPAN) is a good candidate, many other approaches do no support equal capabilities. Assisting this issue, an Intelligent Control Box was proposed by Bai and Huang [51], which works as a *signal converter*. Through a Universal Asynchronous Receiver/Transmitter (UART) interface and an Intel 8051 control panel, the system manages each received packet, and then the Intel 8051 in combination with an Arduino system format the data, in order to convert it into specific commands to control the box. This Intelligent Control Box can manage Wi-Fi-based, ZigBee-based, and Bluetooth-based signals, alleviating significantly the implementation of AmI systems.

Modern technologies have enabled the development of more specific applications. Alharbe *et al.* [52] designed a *Hospital Information Management System* using RFID and ZigBee technologies. They have implemented the concepts of IoT and Cloud computing for patient/equipment/documents detection, locating and tracking. The Hospital Information Management System utilizes the ZigBee technology to communicate mobile patient-aware devices and the main-core system. Through RFID tags identifies objects (individual information, location, and extra data). By providing cloud-computing service enables networking access among the entire system. Each element (separately) of this system is not considered as AmI, but the combination of them follows the policies to belong to an AmI system.

10.4 Conclusions

The IoT is no longer an emerging field of study; however, with almost 20 years of research and development from the research community and large companies, IoT has been finding new ways among modern healthcare. In this chapter, a review of the state of the art of IoT-applications that assist main health issues was discussed. Emphasizing in those that are energy-aware, and implement the newest versions of short-range network's protocols Bluetooth and ZigBee. Although these technologies manage short-range reliable/effective communication, their performance is not reliable enough to be implemented in a real-size healthcare system, which requires error-free activity. That means, further research is required in the re-design of network protocols, in order to achieve better communication performance, and preserve low costs and green energies.

References

[1] World Health Organization (WHO). *10 Facts on the State of Global Health.* [cited 2015 June]; Available from: http://www.who.int/features/factfiles/global_burden/en/.

[2] American Cancer Society, I., *Management's Discussion and Analysis and Financial Statements*. 2014, American Cancer Society, Inc.

[3] World Health Organization (WHO). *Diabetes*. [cited 2015 June]; Available from: http://www.who.int/mediacentre/factsheets/fs312/en/.

[4] World Health Organization (WHO). *Chronic Obstructive Pulmonary Disease (COPD)*. [cited 2015 June]; Available from: http://www.who.int/mediacentre/factsheets/fs315/en/.

[5] Deloitte. *2014 Global Health Care Outlook*. 2014, Deloitte.

[6] International Organization for Standardization (ISO). *ISO/TC 215 Health Informatics*. [cited 2015 June]; Available from: http://www.iso.org/iso/home/standards_development/list_of_iso_technical_committees/iso_technical_committee.htm?commid=54960.

[7] Acampora, G., Cook, D., Rashidi, P., *et al.*, A survey on ambient intelligence in healthcare. *Proceedings of the IEEE*, 2013. **101**(12): 2470–2494.

[8] Waris, M., S.A. Khan, and M.Z. Fakhar, "Factors effecting service oriented architecture implementation," in *Science and Information Conference 2013*. 2013, IEEE: London. pp. 1–8.

[9] Oracle. *Oracle SOA Suite for Healthcare Integration*. Oracle White Paper 2013 [cited 2015 July 8]; Available from: http://www.oracle.com/us/products/middleware/soa/soa-suite-for-healthcare-wp-2046692.pdf.

[10] Roberts, C.M., Radio frequency identification (RFID). *Computers & Security*, 2006. **25**(1): 18–26.

[11] Yao, W., C.-H. Chu, and Z. Li, The adoption and implementation of RFID technologies in healthcare: a literature review. *Journal of Medical Systems*, 2012. **36**(6): 3507–3525.

[12] Rosenbaum, B., Radio frequency identification (RFID) in health care: privacy and security concerns limiting adoption. *Journal of Medical Systems*, 2014. **38**(3): 1–6.

[13] Yick, J., B. Mukherjee, and D. Ghosal, Wireless sensor network survey. *Computer Networks*, 2008. **52**(12): 2292–2330.

[14] Bokare, M. and A. Ralegaonkar, Wireless sensor network. *International Journal of Computer Engineering Science (IJCES)*, 2012. **2**(3): 6.

[15] Mainetti, L., L. Patrono, and A. Vilei. "Evolution of wireless sensor networks towards the Internet of Things: a survey." in *2011 19th International Conference on Software, Telecommunications and Computer Networks (SoftCOM)*. 2011.

[16] ZigBee® Alliance. *ZigBee 3.0: The Foundation for the Internet of Things*. 2014 [cited 2015 July 15]; Available from: http://www.zigbee.org/zigbee-for-developers/zigbee3-0/.

[17] ZigBee® Alliance. *ZigBee Health Care*. [cited 2015 July 15]; Available from: http://www.zigbee.org/zigbee-for-developers/applicationstandards/zigbee-health-care/.

[18] Bluetooth® SIG. *A Look at the Basics of Bluetooth Technology*. 2015 [cited 2015 July 14]; Available from: http://www.bluetooth.com/Pages/Basics.aspx.

[19] Townsend, K., Cufí, C., Davidson, R., *et al.*, *Getting Started with Bluetooth Low Energy*. 2014: O'Reilly Media.

[20] Bluetooth® SIG. *Updated Bluetooth® 4.1 Extends the Foundation of Bluetooth Technology for the Internet of Things*. 2013 [cited 2015 July 14]; Available from: http://www.bluetooth.com/Pages/Press-Releases-Detail.aspx?ItemID=197.

[21] Bluetooth® SIG. *Bluetooth Innovation Training Series Removes Mystery from Developing for IoT*. 2015 [cited 2015 July 14]; Available from: http://www.bluetooth.com/Pages/Press-Releases-Detail.aspx?ItemID=229.

[22] Bluetooth® SIG. *Bluetooth Technology Creates Huge Opportunities in Medical*. 2015 [cited 2015 July 14]; Available from: http://www.bluetooth.com/Pages/Health-Wellness-Market.aspx.

[23] Mendez, E.O. and S. Ren, "Design of cyber-physical interface for automated vital signs reading in electronic medical records systems," in *Proceedings of the IEEE International Conference on Electro/Information Technology (EIT'12)*. 2012.

[24] Rolim, C.O., F.L. Koch, C.B. Westphall, J. Werner, A. Fracalossi, and G.S. Salvador, "A cloud computing solution for patient's data collection in health care institutions," in *Proceedings of the 2nd International Conference on eHealth, Telemedicine, and Social Medicine (eTELEMED'10)*. pp. 95–99, Maarten, The Netherlands, February 2010.

[25] Koufi, V., F. Malamateniou, and G. Vassilacopoulos, "Ubiquitous access to cloud emergency medical services," in *Proceedings of the 10th International Conference on Information Technology and Applications in Biomedicine (ITAB'10)*, Crete, Greece, November 2010.

[26] Poulymenopoulou, M., F. Malamateniou, and G. Vassilacopoulos, "E-EPR: a cloud-based architecture of an electronic emergency patient record," in *Proceedings of the 4th ACM International Conference on PErvasive Technologies Related to Assistive*.

[27] Huang, Q., L. Ye, M. Yu, F. Wu, and R. Liang, "Medical information integration based cloud computing," in *Proceedings of the International Conference on Network Computing and Information Security (NCIS'11)*, pp. 79–83, May 2011.

[28] Jara, A.J., M.A. Zamora, and A.F.G. Skarmeta, An internet of things–based personal device for diabetes therapy management in ambient assisted living (AAL). *Personal and Ubiquitous Computing*, 2011. **15**: 431–440.

[29] Vector. Retrieved on August 2015 from: http://vector-project.com

[30] Keller, H., Juloski, A., Kawano, H., *et al.*, "Method for navigation and control of a magnetically guided capsule endoscope in the human stomach," in *Proc. IEEE Int. Conf. Biomed. Robot. Biomechatron.*, 2012, pp. 859–865.

[31] Caprara, R., K.L. Obstein, G. Scozzarro, *et al.*, A platform for gastric cancer screening in low- and middle-income countries. *IEEE Transactions on Biomedical Engineering*, 2015. **62**(5): 1324–1332.

[32] Jara, A.J., M.A. Zamora, and A.F. Skarmeta, Drug identification and interaction checker based on IoT to minimize adverse drug reactions and improve drug compliance. *Personal and Ubiquitous Computing*, 2012.

[33] Pang, Z., Q. Chen, and L. Zheng, "A pervasive and preventive healthcare solution for medication noncompliance and daily monitoring," in *Proceedings of the 2nd International Symposium on Applied Sciences in Biomedical and Communication Technologies (ISABEL'09)*, November 2009.

[34] López-Nores, M., J.J. Pazos-Arias, J. García-Duque, and Y. Blanco-Fernández, "Monitoring medicine intake in the networked home: the iCabiNET solution," in *Proceedings of the 2nd International Conference on Pervasive Computing Technologies for Healthcare*, pp. 116–117, February 2008.

[35] Centers for Disease Control and Prevention. Retrieved on August 2015 from: http://www.cdc.gov/homeandrecreationalsafety/falls/adultfalls.html

[36] Administration for Community Living. Retrieved on August 2015 from: http://www.aoa.gov/Aging_Statistics/Profile/2011/4.aspx

[37] 2014 published by Alzheimer's disease International (AdI), London.

[38] Díaz-Ramírez, A., F.N. Murrieta, J.A. Atempa, F.A. Bonino, "Non-intrusive tracking of patients with dementia using a wireless sensor network," in *2013 IEEE International Conference on Distributed Computing in Sensor Systems (DCOSS'13)*, pp. 460–465, 20–23 May 2013.

[39] Wang, J., H. Abid, S. Lee, L. Shu, and F. Xia, A secured health care application architecture for cyber-physical systems. *Control Engineering and Applied Informatics*, 2011. **13**(3): 101–108.

[40] Sposaro, F., J. Danielson, and G. Tyson, "iWander: an android application for dementia patients, engineering in medicine and biology society (EMBC)," in *Annual International Conference of the IEEE*, pp. 3875–3878, 2010.

[41] Tabish, R., A.M. Ghaleb, R. Hussein, *et al.*, "A 3G/WiFi-enabled 6LoWPAN-based U-healthcare system for ubiquitous real-time monitoring and data logging," *2014 Middle East Conference on Biomedical Engineering (MECBME'14)*, pp. 277–280, 17–20 February 2014.

[42] Kau, L.-J., and C.-S. Chen, A smart phone-based pocket fall accident detection, positioning, and rescue system. *IEEE Journal of Biomedical and Health Informatics*, 2015. **19**(1): 44–56.

[43] Rakhecha, S. and K. Hsu, "Reliable and secure body fall detection algorithm in a wireless mesh network," in *Proceedings of the 8th International Conference on Body Area Networks* (BodyNets'13). ICST (Institute for Computer Sciences, Social-Informatics and Telecommunications Engineering), ICST, Brussels, Belgium, Belgium, pp. 420–426, 2013.

[44] Dau, H.A., F.D. Salim, A. Song, L. Hedin, and M. Hamilton, "Phone based fall detection by genetic programming," in *Proceedings of the 13th International Conference on Mobile and Ubiquitous Multimedia* (MUM'14). ACM, New York, NY, USA, pp. 256–257, 2014.

[45] Philips. Retrieved on August 2015 from: http://www.research.philips.com/technologies/projects/ami

[46] Cook, D., J. Augusto, and V. Jakkula, Ambient intelligence: technologies, applications, and opportunities. *Personal and Ubiquitous Computing*, 2009. **5**(4): 277–298.

[47] Acampora, G., D.J. Cook, P. Rashidi, and A.V. Vasilakos, A survey on ambient intelligence in healthcare. *Proceedings of the IEEE*, 2013. **101**(12): 2470–2494

[48] Langheinrich, M., "Privacy by design principles of privacy-aware ubiquitous systems," in *UbiComp 2001: Ubiquitous Computing*, vol. 2201. Berlin, Germany: Springer-Verlag, 2001, pp. 273–291, *Series Lecture Notes in Computer Science*.

[49] Wood, A.D., J.A. Stankovic, G. Virone, *et al.*, Context-aware wireless sensor networks for assisted living and residential monitoring. *IEEE Network*, 2008. **22**(4): 26–33.

[50] Lu, C.-H. and L.-C. Fu, Robust location-aware activity recognition using wireless sensor network in an attentive home. *IEEE Transactions on Automation Science and Engineering*, 2009. **6**(4): 598–609.

[51] Bai, Z.Y. and X.Y. Huang, Design and implementation of a cyber physical system for building smart living spaces. *International Journal of Distributed Sensor Networks*, 2012. **2012**: 9.

[52] Alharbe, N., A.S. Atkins, and A.S. Akbari, "Application of ZigBee and RFID Technologies in Healthcare in Conjunction with the Internet of Things," in *Proceedings of International Conference on Advances in Mobile Computing & Multimedia* (MoMM'13). ACM, New York, NY, USA, pp. 5–191, 2013.

Index

www.ingramcontent.com/pod-product-compliance
Lightning Source LLC
LaVergne TN
LVHW011801200926
848484LV00003B/749